The Body Book

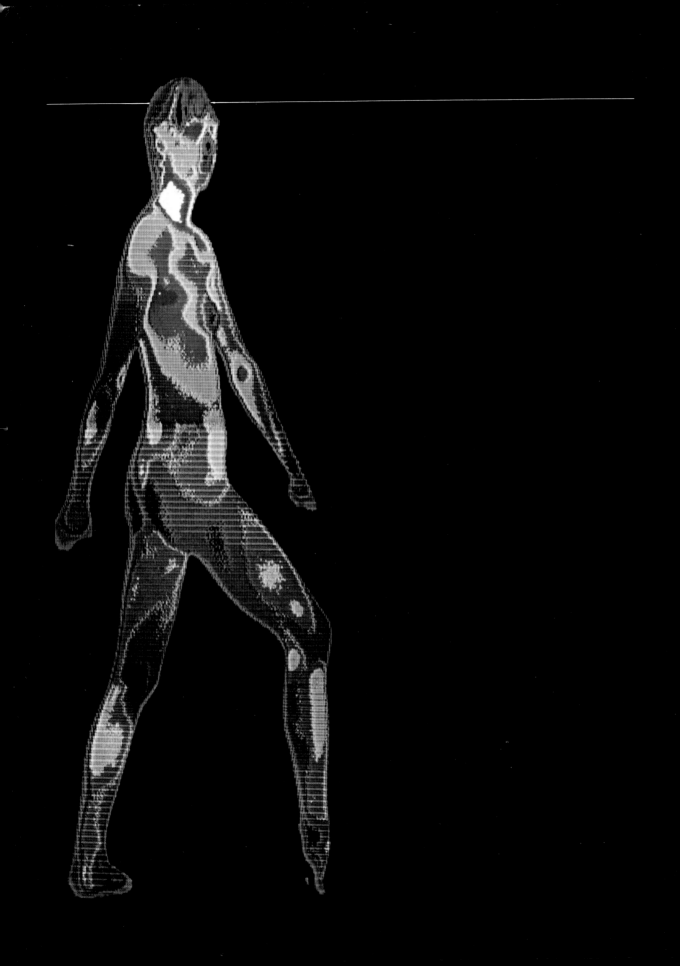

The Body Book

A Fantastic Voyage to the World Within

by David Bodanis

Little, Brown and Company   Boston / Toronto

First Edition

Permissions to use photographs and other illustrations appear on p. 289.

Printed in Japan in 1984 by Dai Nippon Printing Company.
Design and art direction by Dianne Schaefer/Designworks.
Typesetting by DEKR.

LIBRARY OF CONGRESS CATALOGING IN PUBLICATION DATA

Bodanis, David.
    The body book.

    Includes index.
    1. Human biology. I. Title.
QP38.B59 1983   612   84-9737
ISBN 0-316-10071-4

Acknowledgments

*R*esearch for this book could not have been attempted without the excellent computer search facilities of the Centre Nationale de Recherche Scientifique, and the Institut Nationale pour la Santé et la Recherche Médicale. Gracious help was also had from the library staff at both institutions, as well as at the Institut Pasteur, the Collège de France, the Ecole Polytechnique, the Bibliothèque Publique d'Information, the Ecole des Hautes Etudes en Sciences Sociales, the Université de Nice, and the Faculté de Médecine of the University of Paris; and in England at the libraries of the British Museum (Natural History), the Victoria and Albert, and the Guildhall in London, as well as the University of Sussex library, the Bodleian Library in Oxford, and the various libraries of the University of Cambridge.

Further thanks are due: in England, research assistants Dr. Ann Harris (Oxford and the Imperial Cancer Research Fund) and Hugh Tomlinson (Oxford philosopher and now London barrister); Professor J. Z. Young (University College and the Wellcome Institute), for thorough reading of the entire manuscript and welcome advice and discussion; Professor J. A. Edwardson (Medical Research Council Neuroendocrinology Unit); Dr. Saffron Whitehead (St. George's Hospital Medical School); Drs. Dennis W. Lincoln and J. Bancroft (Centre for Reproductive Biology, Edinburgh); Dr. Tony Segal; Professor Sir William Paton (Institute of Pharmacology, Oxford); and kind officials at the Medical Research Council, the British Association, and the Royal College of Surgeons for putting me in touch with other specialist readers. In France, Professor Yehezkel Ben-Ari (Laboratoire de Physiologie Nerveuse, CNRS); Professor Michel Jouvet (Université de Lyon); numerous faculty members at the French Institute for Advanced Studies for free-wheeling lunchtime discussion on this book; Professor Haig Papazian, long of Yale and now most actively retired; Samuel Abt for editorial advice; and a prolific team of typists, Maria Eder, Nicki Holton-Jones, Mary Edwardes, and Esther Bernstein. Also my agent, June Hall, for feats beyond the call of duty; a friendly and professional team at Little, Brown, especially Debra Pearlstein, Barry Lippman, and Mary Tondorf-Dick; and, above all, to Kathleen, for putting up with it all.

All remaining errors of fact or interpretation are of course entirely my own. Kind readers who communicate them to the author will be rewarded with their correction in any future edition.

Introduction

"I *entered biology to understand the beauty of life. I started with*
the ecology of animals, but that wasn't quite right, so I went
one level deeper, to the study of tissues, then to chemistry, and finally
even dabbled in the depths of quantum mechanics. Yet when I got
there, I saw that the wonder of life had somehow slipped through my
fingers along the way." — Albert Szent-Györgyi, *Nobel Laureate, Biology*

This book is most unabashedly designed to catch the wonder of the
human body. It is a look at the extraordinary things that are happening
inside your body every day of your life, and how they affect the way
you feel and act.

The approach betrays the prejudices of a physicist. There is a concern with processes, dynamics, and movements within the body; instead of the usual listing of all bodily processes in a comprehensive
but dry outline form, there is a selective swooping in on the most significant operations of the body. In French terms it's an effort to *démy-*
thifier the body, to place it squarely in the real world we all live in.
Historical events, contrasts with the animal world, and even the little-
known origins of certain common words are brought in with the same
end. It's all to break with the idea that learning about how the body
really works has to be boring.

The pictures have been assembled to bring out the extraordinary
dynamics inside the body during ordinary emotions such as worry or
love and commonplace actions such as standing up or having a drink.
Most of the illustrations are from European sources; several have
never been published before.

Dancing strands of active heart
wall muscle. Stress can make
them move more quickly, which
translates as your heart beating
faster.

Turbulence in air produced by a moving object, such as a moving hand.

Contents

The Body Book

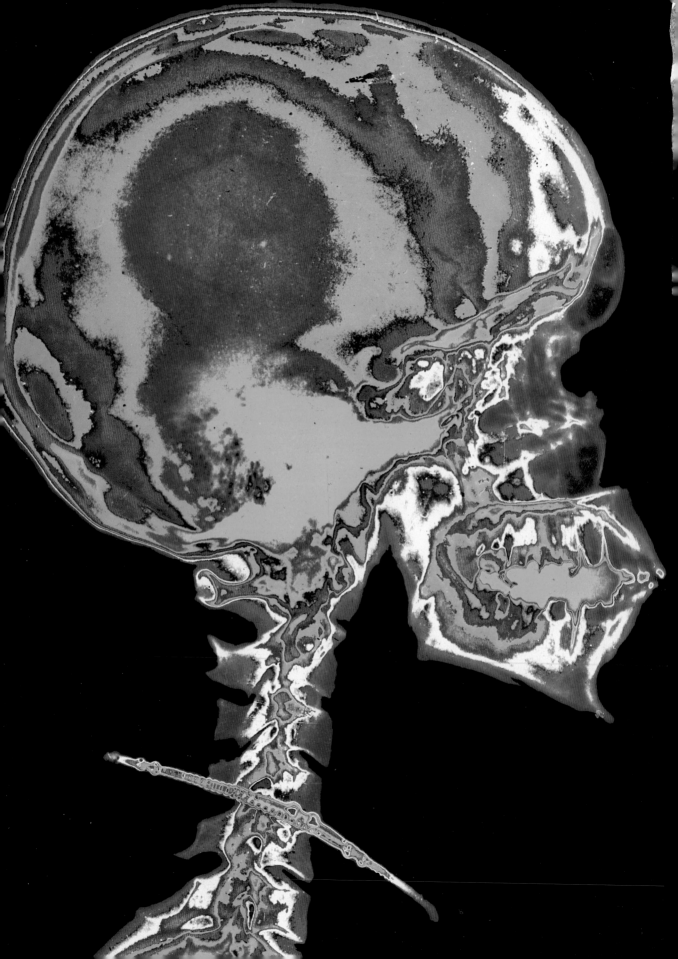

1
Activity: The Day Begins

*E*ach time a jogger takes a step forward, his leg bones take on the
stress lines of a Gothic cathedral exposed to high winds, his eyeballs bounce slightly in their sockets, and his heart slams backward
like a football being kicked by a heavy boot.

Even minor movements during the day, such as poking an elevator
button, or slumping down in a chair with a book, do a tremendous
number of unsuspected things to the inside of your body.

There's no way such changes can go on without affecting the way
you act, the very way you feel and think. This chapter is about how that
happens. An ordinary routine of a typical morning shows them all.

7:00 Getting out of Bed

There might be some people who try to get out of bed by tossing their
legs high and keeping their heads down, but for most of us it's more
usual to keep our legs down and just raise our heads. This seems to be
the simplest movement possible, but there's a lot more going on than
there seems.

Consider the moment when your head is on the pillow, you've digested the fact that the alarm has rung, and after a certain amount of
time meditating on the warmth and cosiness of the bed, have finally
decided to stop malingering and get up. Up lifts the tousled head with
one quick start, which is fine, but somewhat less quickly moves the
brain within that head, which is less fine.

That's because the brain is not wedged solidly inside the skull, but
is instead dangling about rather loosely, like an artichoke perched on
its stalk, on the end of a long stem sticking up from the spinal cord. As

*Color skull, suggesting stress
lines along which cranium
would react to sudden acceleration. Note subject's necklace.*

you start to sit up in bed the skull moves up quite nicely, but the brain, which hasn't really been informed what's going on, and has no muscles attached to it to help if it did know, tends by inertia to stay still where it was.

That sets the scene for a nasty collision, because the skull is bone hard and somewhat ridged, while the brain, however·much pride we have in our own, is a soft and sodden object made up 85 percent of ordinary water. If the skull of our sitting up individual did collide straight into the back of the brain, it would not be the skull that would suffer. Prelaid stress patterns in the skull would carry the strain safely away. Rather, whole globs would slough off the back of the brain and we would certainly not be welcomed into the office at 9 A.M. staring blankly as one is likely to do after having just had a few crucial layers containing millions of brain cells so removed.

Such a fate is spared us, since the brain is ingeniously cushioned from banging into the skull's inner surface. The brain never gets in direct contact with the skull, but instead stays wrapped in what look to be three very thin, and very strong, plastic bags.

Between the first two of these is an innocuous looking fluid, called CSF (for cerebrospinal fluid). It acts like a car's shock absorbers. As the back of the skull comes up from its resting place on the pillow, and begins to move in on the unsuspecting brain, it squishes in on the CSF in its slender bag. The CSF in turn splays out and squishes its surrounding bag up against the back of the brain. It's a lot gentler than the inside surface of the skull thwacking directly would have been. The fluid-filled sac eases the brain forward, in the direction the skull is going, without doing it any harm: the ultimate antiwhiplash headrest.

As our bleary-eyed subject continues to move his head upward, the cunningly placed fluid keeps on getting pressed in, and so keeps pushing the brain safely away from the potentially sharp and abrading skull.

Finally when the sitting up is completed, the fluid from behind the brain stops being pressed in and is free to swirl gently around the sides of the brain from back to front, there to bring the brain swinging forward to a gentle and unscathed halt, this time protecting it against the inside *front* of the skull.

Nor does the fluid go away when the difficult maneuver of getting out of bed has been finished. Every time the head moves during the day, twisting to the side to look out of a car window, or nodding vigorously forward in the presence of the boss, the cushioning fluid is not far behind, sloshing from side to side in just the right way to keep the brain cells where they belong, on the brain, and not embedded in microscopically jagged crevices in the skull.

There are limits, of course. Turn your head too fast and it hurts, for the fluid will barely have time to do its cushioning job before the skullbone slams into the brain. Go faster and it's even more serious. Any acceleration of the head of more than about 35 mph per second is likely to cause unconsciousness, and as all boxing referees know, a sharp blow to the back of the neck, a rabbit punch, can do this with high efficiency. That's why bouncers are prone to use them, and that's why in the ring they're illegal.

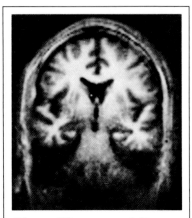

Like cauliflower on a stalk: scan from behind of living brain perched in head. The whole will oscillate back and forth, frondlike, when acceleration is applied to surrounding cranium. Note treelike branchings in two structures slightly below midline left and right — they're fat-covered nerve impulse pathways in the cerebellum, which controls muscular coordination and sitting up. Image produced by measuring resonance of atomic nuclei in the brain subjected to powerful magnetic fields.

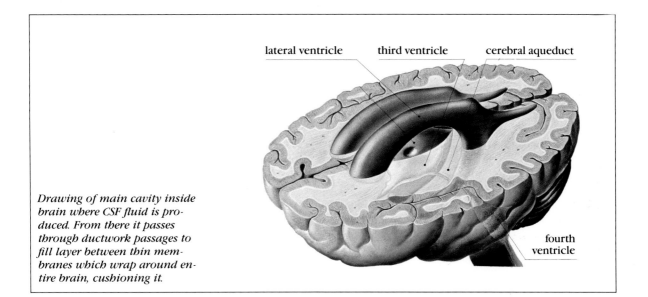

lateral ventricle   third ventricle   cerebral aqueduct

fourth ventricle

Drawing of main cavity inside brain where CSF fluid is produced. From there it passes through ductwork passages to fill layer between thin membranes which wrap around entire brain, cushioning it.

Unfortunately enough legal swings to the jaw can have the same effect, when a tough-punching fighter is behind them. The autopsied brains found in the skulls of boxers who died in the ring after a long career of such battering almost invariably have a thick layer torn off all around. The inner remains have been variously described as looking like toothpaste, or guacamole sauce — a sticky finish indeed.

The buffering fluid that usually spares us such a fate sounds too convenient to be true. What wonder fluid is it that could do all this? Where does it come from? The answer is somewhat less appealing than one might hope. For the fluid comes from big holes inside your brain, and contains among other substances the suspiciously named urea (from the Greek *ouron* for urine), which is in fact one of the major constituents of that other fluid that is usually stored in our bladder. This is the real nature of the safety mechanism that allows us to move our heads with impunity and safely get out of bed in the morning.

Still the holes in the brain should really be nothing to feel ashamed about, for everyone has them. They're leftovers from the time when early as fetuses each of us had a long hollow tube going along our back and into the head. Later in the fetal stage, nerve tissue grew around the top part of the tube to form the brain, but never quite filled it in. While the brain grows after birth the hole inside it grows too, and splits sideways so that one half ends up in each side of the brain. By the age at which most people have their driving license each hole has reached its full size almost large enough for a penny to fit in.

That's where the CSF is formed. It drips out from the top of these holes, filtering across from thin blood vessels in the brain tissue just beyond. A small half-glassful dribbles in every day, and sloshes around a bit before spiraling down something resembling a narrow drain. That leads out into the space between the two innermost of the thin sacs that wrap tightly around the brain, and that keeps the CSF shock absorber pumped full and ready. In creatures less given to ratiocination

than ourselves, such as the gentle kangaroo, there is more hole
and less brain, and the drain itself widens to create two more holes,
Gruyère cheese like, just as big as the original ones.

CSF does other good work on top of absorbing shocks. The urea it
contains is there because CSF is a convenient way of floating waste
products out of the brain; it would be unseemly if we had to interrupt
our conversations to have the brain get rid of these unwanted materials itself — perhaps with a sudden side twitch of the head to send it
out the ear. Urea is over 50 percent nitrogen by weight, and is the
body's chief way of excreting that easily overaccumulated element.
Sweat, nails, hair and skin flakes are other ways it gets unloaded. Since
the CSF will eventually pass back into the bloodstream through the
outer layer of the saclike membranes in which it's enclosed, that's a
simpler way for the urea to be sped on its recommended way. The
flow is always an even one, for unlike blood, CSF does not clot. It's
clear and as slippery as water, easily carrying dissolved gases and electrically charged particles on its outward way too. Vampires would
hate it.

Our CSF isn't just a passive carrier either. If some unwanted bleeding inside the brain looks like causing a dangerous rise in pressure,
the holes in the brain that produce the CSF in the first place will cut
down their output to keep the overall pressure constant. (The aqueous
humor in front of the lens of your eye balances its own pressure in the
same sort of way.) It won't be enough for a major gusher, but will
keep things steady while any of the hair-thin scratches we so frequently get in the brain go about healing themselves. It can keep what
is usually an enveloping safety cushion from turning into a startling
pressure band.

Knowing the peculiar way in which the brain is cushioned against
sudden shocks gives another insight into getting out of bed in the
morning. So many people resist getting out from under the covers,
and putter about in a sullen mood acting grumpy over coffee once
they do. Why?

One explanation is precisely because of the fact that our brain is
cushioned by this crucial CSF, for the cushioning only works up to a
certain speed of movement. Get up too fast, and the brain gets banged
too much. It hurts. That makes a distressing contrast with the wonderful world of dreams, where we ordinary mortals become superheroes,
unfettered by the bounds of rationality and polite behavior, let alone
gravity and the ache one gets from moving too fast. While asleep, one
is infinite and free to soar; while awake, bound to the finite and restrictive. Lifting up the head makes that only too clear.

The power we have in the dreamworld would be continued in the
day only if through some change we could leap powerfully out of bed
in a single bound without drudgingly repositioning our feet, and carefully standing up. Alas, even if we had the muscles for this, the acceleration produced in such a bound from the supine to the vertical could
be properly cushioned only by a much thicker envelope of CSF
around the brain than we now have, and to keep the same brain size,
that would mean a much larger skull. Since the cranium babies are

born with even now is just as large as it can be still to fit through the mother's pelvic girdle that puts a limit on how quickly and powerfully we can leap out of bed. And that, in turn, hardens the line between the life of dreams, and the life or ordinary day.

7:01 Standing Up

Despite the reproaches, most people do decide to get up. Then they do a most remarkable thing: they stand. A number of animals try but haven't quite got the idea. Crocodiles, for example, have stubby little legs splayed out to the side, and rest their bodies on a sheet of muscle that's strung across their bellies from one leg fixture to another. Their difficulty comes because they're much closer to the first crawling creatures — fish that lived in mud flats and pulled themselves along commando-style on their side fins during the dry season — than we are.

The more advanced West African gorilla does a bit better, being temporarily able to heave its 600-pound bulk straight up and bellow a war cry at any intruder in range. But within a few seconds it's ignominiously forced to plop back down to earth and lean forward on its knuckles because of the pain it gets from standing straight.

That raises the big question that everyone who's ever had lower back pain in the morning wonders about: Is standing something we're designed for, or is it a developmental quirk that our bodies haven't yet worked out? The answer is that it's both, an incredible hodgepodge of mixed strengths and weaknesses.

The advantages begin with a design that really is specialized for standing up. The poor gorilla is like most four-legged creatures in having an almost straight column of backbones topped by a slight pile of curving ones. That straight column is an excellent suspension bridge from which the liver and other internal organs can dangle, while the curving ones on top allow for that bit of flexible tubing below the head that we call a neck. But it does transmit all the stresses of a standing animal smack through the small of the back, and that's what makes their standing up for long impossible.

Humans *are* different. We are born with a relatively straight spinal column, but after about 13 months the bottom few bones of the back stop forming a straight line and begin to curve backward. (Cartilage is what does this — stuff quite like the filling of your outer ears that makes them flappy.) That svelte curve makes the back bones into just enough of a spring to turn what would otherwise be overbalanced and painfully jarring steps into the natural bounce of ordinary striding. When the change occurs, infants walk happily tottering proof of this key evolutionary addition. The cartilage in there that helps do it is similar.

It's this property of having a lot of small bones perched at a springy angle one on top of the other, instead of just one massive rod, that allows for all the turning and bending that are seen to perfection in ballet. Ancient Romans recognized this arrangement and termed the backbone the "turning" column, a phrase that Latin influences into *vertebral* column — the name still used in school texts today. *Vertigo* comes from the same root.

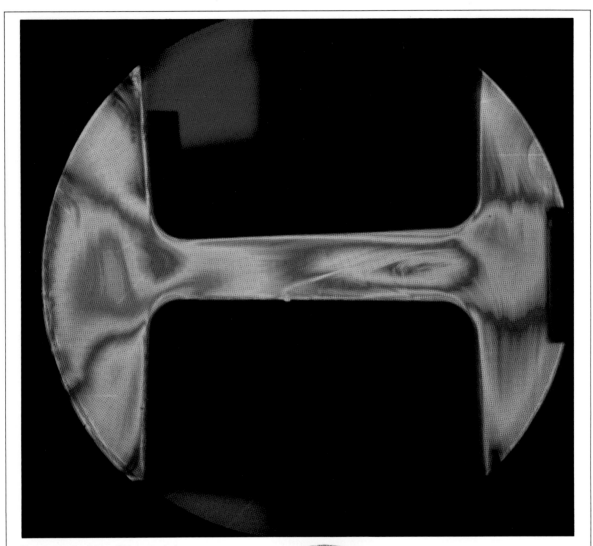

Pressures inside solid material revealed with polarized photography. Bunching of lines indicates highest stress, especially in outside of long beam, and outward curves at top and bottom. The thigh bone has to support the same such vertical stresses, which is why it's thickened on the outside of its main tube, and rounds outward at top and bottom.

Below, top of thigh bone cut open to show bone growth matching stress patterns revealed in interference photograph.

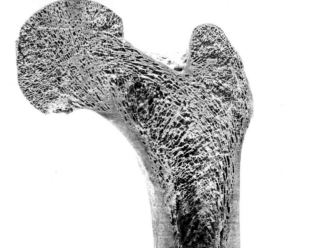

Standing high does more. It frees the hands for more interesting tasks than just supporting the body's weight. It also pushes the horizon farther away: a six-foot man standing on a plain can see to a horizon that's about three miles away; crouched down on all fours, it would be only one mile away. In primitive times, and in certain unwelcoming city neighborhoods today, there could be many miscreants one would wish to know about before they got in that closer range. (It's also a curious fact that you're always taller first thing in the morning than at other times. That's because your backbones stretched out when you were horizontal sleeping, and gravity wasn't compressing your spine then either.)

The disadvantages due to standing however, are all consequences of the main, backbone-curving change. One is that if our early riser leans way forward to tug on something, as in stretching across the recently relinquished bed to pull up a sheet caught under the far edge of the mattress the advantage of the bend in the spine is neutralized by the reverse bend of the lean forward. In that precarious position, any tugging or pulling will not flow neatly down the spine. Instead it will be routed straight against it, at right angles, and however strong the back bones and tension members are that's far from fair. Even a massive concrete dam can be knocked over by a fairly small force pushing into it at the wrong angle, as happened with the special water-hopping mines the RAF launched against the easily stress-fractured inner side of the Möhne and Eder dams in Nazi Germany in 1943. The mines merely chipped tiny fissures into the flat side of the dam, but that was enough. The water that got in those chips pushed against the dams at an angle it had no defense against, swelling the tiny fissures into splits that went all the way through until the gigantic dams burst apart.

For someone stretched out making a bed, the consequences are similar, though on a less dramatic scale. The lines of force pulling at an abrupt angle to the spine can actually knock one of the ringlike spinal bones out of position, or as happens more frequently, squish and deform some of the jellylike contents of the cushioning disks between these bones for impetuous old people or anyone who's not particularly supple. It's an ongoing hazard. At least none of us are in the precarious position of certain lizards who have a weak spot where they can actually sever their own vertebral column with one mighty muscular contraction.

Another unfortunate consequence of our standing position comes from the curious weakness of the muscles that cover the abdomen. They're usually too feeble to stay taut much past adolescence, and sag out in the cruel yet far too aptly named "potbelly." It's by no means a uniquely modern affliction. English peasants of well before Shakespeare's time knew a good joke when they saw one, and started calling protuberant stomachs after their word for the huge, swelling bellows they used to heat fires for metalworking. *Belwes* for them, *belly* to us.

Getting a potbelly always happens in two steps. The muscle fibers in the abdomen just under the surface are built of very peculiar molecule chains. One type is straight and relatively broad; the other is braided and relatively thin. The two sorts stretch against each other, in

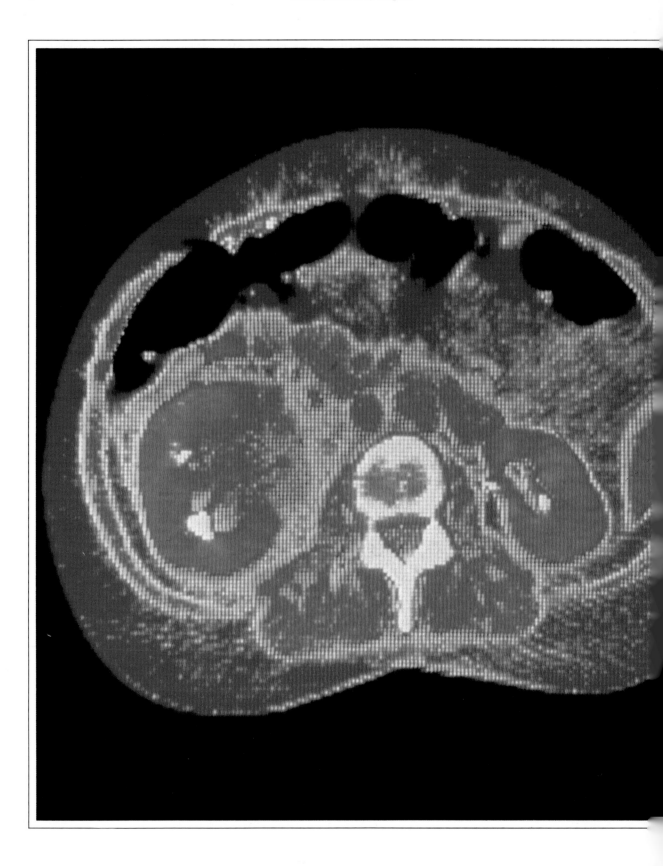

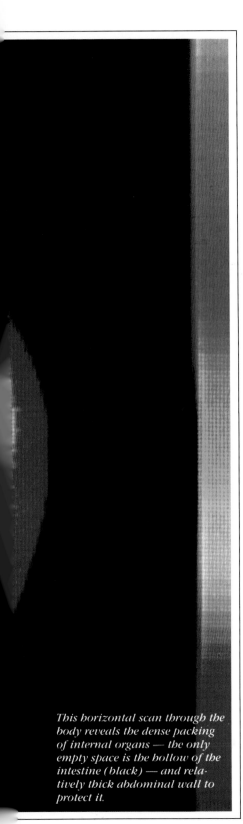

This horizontal scan through the body reveals the dense packing of internal organs — the only empty space is the hollow of the intestine (black) — and relatively thick abdominal wall to protect it.

a repeated formation that can pack many trillion in just the first inch above your navel.

Sticking out from the thicker chain is something like a fine and sticky Velcro fur. Normally it just sits there and doesn't do anything. But when muscles get the signal to tense, the fur from each thick chain reaches out and pulls the neighboring thinner chain past the thicker one. The result is a shortening of the entire muscle fiber, which we feel as a tightening or flexing of the abdominal wall. With increasing age, however, the entwining and contracting doesn't work as well as it used to.

That's trouble. For just at the age when the filaments begin to loosen — somewhere around the mid-twenties or so — a midriff extending force is likely to be surging out against them. This is a visitation of the omentum, a spongy membrane which is especially full of fat globules and drapes apronlike down the inside of the abdominal wall. This enlarging sheet pushes out on the thin and weakening abdominal wall, and a noticeable potbelly is the result. Rare are those adults who can stand sideways in front of a mirror and breathe outward without noticing the pudgy protrusion swell forth in all its useless bulk.

But enough of such unwelcome aftereffects. The worth of standing need not be judged just on the functional question of efficiency in stature or movement. Its interesting just to know that your head shrinks in volume when you stand, due to blood being pulled by gravity out of it, and your leg swells in volume, due to some of the same blood being pulled by gravity into it. There's also a rarely seen aesthetic side to standing and it's due entirely to the body's warmth.

Heat given off by a standing body spreads steadily up. It turns out that by the ankles of every naked standing adult there's a current of rising warmed air that is over a half-inch thick. By the waist this upward gust fills an invisible inner tube five inches thick around the body, while at the level of the head it can be eight inches thick, and rising strong. Clothes or a bathrobe will reduce the spreading, but not the inexorable upward blast. Shirts with tight collars produce more heating near the shoulders than the middle of the back precisely because they trap this rising column.

Most curiously, this rising air jet is tapped into as an aerial transport system. What it transports is most of what we know as household dust. More than 80 percent of this dust, which one can see floating when a sunbeam shines across the room, is made up of little pieces of skin. Millions of hard, dead skin cells get scraped off even the cleanest adult body when its owner puts on a shirt in the morning or slips into a bathrobe. The rocking of the chest from breathing sends hordes of these dusty messengers falling off, as does any brushing of the arms against the side, brushing of the fingers against each other, or aimless touching of the body.

Ten billion fragments come off this way every day. Most of them get caught in the upward river of heated air around the body, and up they go, way over the head, there to form something like a tenuous mushroom cloud before spreading out and settling down on any surface

they should find. Over forty pounds — a large suitcase worth — of
dead skin cells float away this way in a lifetime.

When doing a bit of dusting before the guests arrive, you're really
quite fittingly dusting away a bit of your standing self.

7:01:15 *Walking*

The mighty creature who has surmounted so many evolutionary obstacles to stand by his hearth in the morning, our ordinary commuter
standing groggily awake by the side of his bed in his pajamas, is now
ready to step forth. He does so in the most peculiar mode of locomotion, walking.

Just like standing, almost no other creature does it. There is trot-
ting, slithering, swimming, crawling, and hopping, but not much walking. It's an inherently unsteady movement, involving as it does the re-
lentless attempt to balance first on one leg, then the other, in what is
little better than a controlled fall.

The whole balancing job has to take place on an area just a few
inches square — the bottom of the feet — and not even all that, as a
good quarter of the foot doesn't touch the ground but instead curves
upward like the arch in a masonry bridge to help take the shocks of
each step.

No architect would stand for it. Skyscraper designers automatically
make the bases of their buildings at least as wide as the greatest width
on top; some are not content unless they can have them splay even
farther outward. For greater support anchoring columns are always
sunk deep into the earth too. Now a base that wide for a human would
be a wedge of foot as wide across as the hips. Any skyscraper perched
on two points as relatively narrow as the feet would come tumbling
down with even the slightest breeze. As we'll see, walking takes about
two hundred muscles, adjusting this strut here, pulling that section
there, just to keep the upright human body up.

Walking also produces even more bizarre special effects than the
rising airstream of standing did. Many of these effects are only visible
to the unaided eye in extreme situations, such as the great acceleration
of a rocket sledge or a high-speed parachute jump. However, they can
still be found, in appropriately reduced form, in ordinary movement.
First, during walking, is that the cheeks of the face flap. Now, cheeks
are just there for such light-powered tasks as keeping food in the
mouth from dribbling out through the sides, or allowing useful signals
such as the grimace, smile, and wink to be conveyed. It shouldn't be
surprising that they don't have much muscular bracing against flopping outward with each downward motion of a step forward. It's not a
great flopping, but it makes ultra-slow-motion films of a walker look
like a scuba diver ostentatiously filling his cheeks wide with air and
then letting it out. Anyone who walks with deep bounces in each step
can sometimes feel it for himself.

Even those who think that they don't bounce, and pride themselves
on a stately glide, are in fact oscillating all over the place. The eye, for
example, which is only about 6.5 cc in volume, rests in a cavity in the
skull that's a loose-fitting 29 cc. It needs extra help to keep from fling-

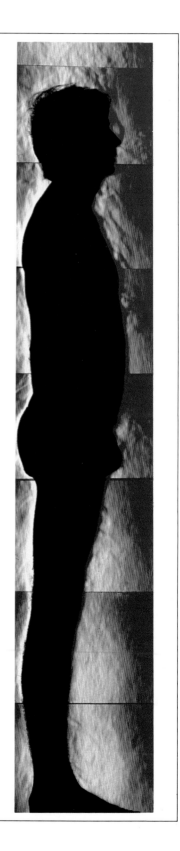

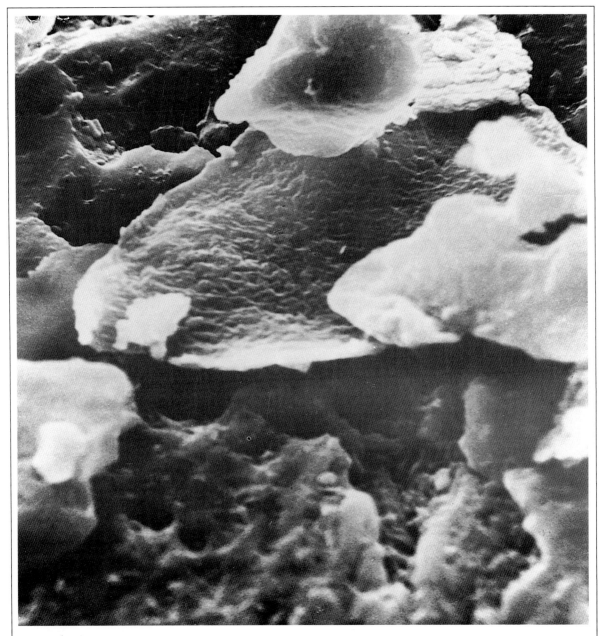

*Hardened skin cells, some of the
ten billion that slough off even a
clean body each day.*

*A heat-flow photograph of rising
air currents produced by a
standing body.* ◄

ing up and down in time with each step, and fling, to a definite extent, it does.

Other bouncers are the internal organs such as the stomach, kidneys, liver, and heart, held in place as they are only by their shapes, the blood vessels that supply them, and the thin, all-enveloping lining sheet of the abdomen — the peritoneum. Also the hairs in the nose, the skin on the elbows, the jawbone, the whole head on the neck, the outside of the ears (this being the one time when everybody can wiggle their ears), folds of fat on the stomach and upper arms, the buttocks, lips, both facial and labial, the scrotum, the vocal cords, the bladder, the teeth, the toenails, and, in those so equipped, the appendix and tonsils.

Quite a spectacle, and no wonder a journey in a car with hard shock absorbers can produce the feeling that you've been shaken inside out. You have.

Most of this shaking in an ordinary walk can be seen only with special cameras. A noticeable exception is the scrotum, for like all good pendulums it swings faster with increasing length a law first discovered by observations on swaying chandeliers at the Cathedral of Pisa in the 1580s. In hot weather, or after emerging from a hot shower, scrotal length increases to make enough of a swing so that few men will go striding off in those conditions without some helpful cotton or nylon bracing.

Breasts bounce too, with the nipple always bouncing the most, and although in most cases not unpleasant to the possessor or beholder, attempts to restrict its oscillations have included woven leaves, strings, binding cloths, whalebones and struts, binding elastics, and, in a change first popularized by the Parisian corsetière Hermine Cadolle just before World War I, that cross-braced stalwart of stress redirectors, the bra.

Hardly any of this bouncing is quiet. The creaking of bones is well known, but a well-placed sound boom, rigged for vibrations of a lower frequency than the ear can detect, would be able during our early riser's shuffle across the bedroom floor to hear and perhaps later process and play back a totally distressing mix of sloshing fat, rumbling cheeks, squelching eyeballs, jouncing skin, and quivering cartilage. Sometimes the sound is high frequency instead of low. In the first step of the day there's a big spurt of blood gushing up from the legs, as all the fluid that build up down there gets powerfully cleared. Ultrasonic dopples shift detectors strapped onto volunteers' legs pick it up as a squeak.

Clothes can add to the chorus too, with their shaking over the skin. The rule is that the faster the shake, the louder the clothes-initiated noise: a karate specialist's punch can cause a snap like a whip cracking when the material of his jacket sleeve flies forward and is just as quickly pulled back. For even greater speeds, it has been computed that the two passengers sucked out of a broken-open DC-10 shortly before it crashed near Paris in 1974 had to listen to a noise like artillery shells going off, produced as their clothes vibrated in the four-hundred-mile-an-hour wind.

The surprisingly large bony pit known as the eye socket. The six ocular muscles fitting around the eye can turn it quickly when tracking an object, but are not secure enough to hold down all its bouncing in the course of movement. The eyeball fills less than half this volume.

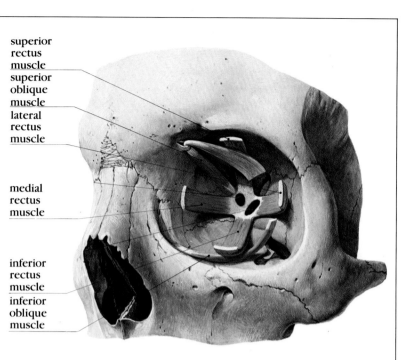

superior
rectus
muscle

superior
oblique
muscle

lateral
rectus
muscle

medial
rectus
muscle

inferior
rectus
muscle

inferior
oblique
muscle

Depth contours of a woman's breast. At rest they're oval. In the peak of a running step, inertia and lack of restraining musculature will deform the contours to circles. Their return back to ovals at bottom of step is what appears as full bouncing cycle.

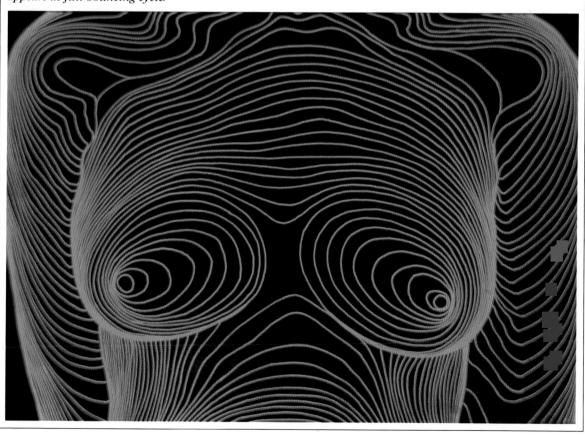

The clatter of the body making its first steps from the bed in the morning would not continue for long without the help of two tiny organs, buried deep in the skull behind the ears, which work by measuring the sloshing of fluids.

The first is a pair of sacs that manage to measure how much the head is tilting. It's a meritorious task. Imagine someone who tilted his head and had no organ of equilibrium to register it. As his head tilted down toward his shoulder, all he would see would be a shoulder getting larger. Looking out at a room with the same unnoticed tilt of the head would be no less disconcerting. It would seem as if the whole room had begun to rotate.

We are spared such problems by the two tiny bags, wedged just next to the hearing sensor of the inner ear. Within each bag are thousands of even tinier boulder-shaped crystals, that are submerged in a clear fluid, and resting on an uneven surface.

As the sac turns in a shifting head, the uneven surface turns with it, but the relatively heavy crystals tumble backward trying to stay in the same place they were in before the turn. With each tumble they land on different nerve fibers, just hard enough to jar them into sending nerve signals to the brain.

Both the signals and the tumbling depend on the amount of tilting the head is doing, so the result is a constant readout to the brain of how the head is tilting. Because the crystals are so small — it would take over a dozen to equal a salt crystal — they measure the head's tilting to a wondrous degree: most people can feel changes in their heads' angling of as little as a few widths of a hair.

The only inaccuracy no one can get away from is a rather tiny time delay. Lean your head forward as you're reading this and you'll seem to have the sensation of being closer at the same time as you do the leaning. That's misleading though, for the crystals won't have finished tumbling, and so the brain won't get word of quite where the head is, until a bit after the move has been made.

As a delay it doesn't make much practical difference, but it does raise the curious philosophical notion that we perpetually live in a world of perceptions that are just slightly out of kilter with the real world. Next time you look at a friend, a moving car, or your own hand, reflect that the image you see is not an accurate picture of your hand at the time when you are doing the seeing, but a delayed photo memory, delayed by the time it takes for the light to reach your eyes, get processed, and travel into the brain.

The delay is slight, less than a thousandth of a second, but it is there, and everything we see, hear, or sense from outside is already just a memory of a now-vanished frozen tableau. At every instant we are deaf, blind and senseless to events as they happen, as is everyone around us.

The other balance organ in the inner ear is designed to measure any spinning or turning the head might be going through. It looks a bit like a hollow three-ringed pretzel, and is made up of three hollow tubes jammed in at right angles to each other, each pumped full of a fluid as smooth as ordinary water.

Whenever the head starts to move, the fluid stays still in its tubes for a moment, out of inertia, and so tugs a bit on flaps attached to the ends of delicate nerve endings on the inside of each of the tubes. That tugging sends a signal to the brain, through a convenient hole in the skull for the nerve fibers to pass through, registering that some movement has started.

These signals stop when the fluid begins to move in time with the hollow rings, and so stops tugging on the nerve ends. Only when there's another sudden change that leaves the fluid whiling away behind will they start up again. It's a great way to know how the head is moving. It also makes sure the eyes move, too, so that the world doesn't go around when we turn the head.

Sometimes, however, it's too great, and sends along position readings that the owner of the head would be glad to do without. This is an all too common occurrence aboard a ship, or in a plane that's gotten too close to a thunderstorm. In either instance one awful lurch in position follows another, which the pretzellike structure dutifully sends along to the brain.

That's where the trouble begins. Each time a ship begins to pitch upward, for example, the vigilant mini-pretzels let the brain know it. Through nerve lines not under conscious control the so-informed brain squeezes the muscles around the stomach, or turns the eyes in the direction of the pitch, to try to adjust to an upward moving ship.

If matters stopped there it would be OK, but the pretzel protectors are not as lax as that. When the ship starts to slam down from the wave's crest the fluid rings are eager to inform the brain that there's been a change in plans. Amiable idiots. Their message makes the brain reverse its instructions to the stomach, eyes, and the like. This happens again and again, with each wave, even though the passenger at the center of these automatic ministrations might be desperately aware that the ship is of course going to continue to go up and down, and that it's ridiculous to tense and untense the stomach each time it changes direction just because the tiny balance measuring pretzels say so.

The ancient Greeks, who sent colonies in ships all through the storm-ridden eastern Mediterranean, had just as moronic balance detectors in their ears as we do today. It's little wonder that they called the particularly distasteful feeling that this constant changing of orders produces in the stomach after their word for ship. That was *naus,* as in the innocuous *nautical,* whence the feeling we label *nausea.*

Such overcompensation is the exception, however, and for most purposes these balance centers are indispensable. People in whom they are damaged describe life as taking place on fantastically slippery ice skates.

All creatures with backbones have such balance centers, even those like the turtle, which have no outer ear and make do in a silent world. They must still know how they're moving. The sacs, boulders, and pretzel rings in our inner ears are so estimable, in fact, that they are imitated by those public-spirited physicists who spend their days building the missiles on which nuclear warheads are to be carried. For such men, the steering of their creations is a problem, as even remote control by inbuilt TV cameras is vulnerable to radio jamming.

U.S. nuclear missile specialists solve this problem with a system of tiny metal spheres, which are held in place by wires or electromagnets, locked in a box, and hooked up to the steering controls. These spheres tumble with each acceleration and twist of the rocket, just like the fluids or small crystals in the inner ear. If a blast of wind in the stratosphere knocks the rocket to one side, or uncharted gravitational perturbations jerk it suddenly downward, the lightly held spheres will begin a roll ever so gently to the other side of their box. Deft electronic sensors measure and use this motion to correct the rocket's position accordingly. It's the same as if someone had stumbled suddenly to one side, and then caught his balance when his brain registered the swirling of the fluids in his inner ear. Hopefully it will not be as accurate.

For the body wavering there by the bed, chock-full of wedged-in balance centers, all that's needed to produce walking, steady or otherwise, is for the muscles to start doing their stuff. A lot of muscles. Most people know of the biceps in the arm, possibly the triceps and deltoid, and have a feeling that there are another dozen somewhere else. But in fact it takes probably two hundred or more distinct muscles to get anything like a single walking stride to take place.

Just getting the right leg to lift forward takes over forty muscles tugging in unison, like deckhands hauling out a boom. Once those forty muscles have started a ponderous step though, a lot has to happen before their opposite numbers on the left side get a chance to work. For a body that has just swung one pajama-clad leg forward is in imminent danger of doubling over forward — remember the straw scarecrow in the Wizard of Oz? — as the momentum of the half-step works its way through the torso. Muscles in the back are quickly needed to pull in on the shoulders and chest, to return them to a straight-up stance. Pull on them they do.

Yet by themselves even those back muscles would do more damage than they save, for by stopping the forward movement they start the abdomen, chest, and head leaning backward, a motion which if unhampered would produce a most undignified plop upon the buttocks. To stop *that* overcompensation yet another series of muscles joins in, the ones stretching up the abdomen. Someone frantically trying to keep from falling on slippery ice is likely to be demonstrating, much against their will, all these overdone moves to perfection. Only when they've all done in the right amount does a vertical position result. To help in the extraordinary control needed for walking you even have little cigar-shaped organs — internal eyes they've been called — sprinkled thoroughly all over your muscles, that send back useful readouts to your brain or spinal cord on how that muscle is moving.

Once the straight pulls are worked out properly it's time to start on the twists. The first step out with the right foot started the body twisting around its central axis, which would be fine as a first step in the judo throw osoto-gari, but which must be countered by sideways-pulling muscles in the hips, chest, and stomach to keep that step away

Weight distribution problems of world class runners are just ordinary ones magnified. Note feet off ground, torsos overbalanced forward, extreme hand displacement from trunk.

Muscle speed champion — a leopard frog taking off. When walking we continually balance ourselves by an unconscious tensing and relaxing of dozens of different muscles in a very precise time sequence. The fastest nerve signals controlling this go down broad nerves that have regularly spaced fatty coverings on them, so allowing the nerve signal to "leap" from one gap in the covering to another. Frogs have a different mix of nerves, including more very broad, fat-covered ones, which allows them to control their supporting muscles fast enough for moves impossible to us. Possibly natural athletes have more of those special nerves too.

from ending up like the judo move. That takes more muscles and that's why the first steps in the morning account for the two hundred pulling muscles. (Lest this seem insurmountably trying for the final thing in the morning, note that just opening the mouth to grunt "hello" means operating the muscles of your lips, jaw, tongue, palate, pharynx, larynx and respiratory system in an intricately coordinated program that requires over five hundred contractions per second.)

These locomotory stalwarts are not all as equally appreciated as they should be. Among them is the widely respected long muscle of the outer thigh, which under normal conditions is one of the strongest in the body. (The uterus during childbirth can tense almost as powerfully as any — a rare, and brief, exception.) But among them is also the rather less honored gluteus maximus, which, neatly placed behind the hips, above the thighs, and below the waist, is what forms those hemispheres upon which we sit. The muscle has a key role in pulling the body up on our legs every time we stand.

The buttocks also seem to grow much too easily. However, most of what is noticed in that region is usually just the fat layer covering that mighty workhorse, not the muscle itself. The fat layer is especially common in women, and despite a wealth of derision seems due more to the spongelike properties of the surface tissue in that area than to any moral lacking of the afflicted party.

As consolation there are certain human groups, most noticeably the Hottentot peoples of southern Africa, who have developed fat storage in this area to truly epic proportions. Numerous cooking utensils and sometimes a small child can be placed on these women's mighty backing. It is an excellent adaptation in a climate where food is often hard to come by and seems to have only the most pleasant of effects on the surrounding Hottentot men. But in western countries such relativity of beauty is less frequently accepted, and we are at lengths to disguise even mention of this noble muscle. The official attempt at concealment, *gluteus maximus,* is but an archaic rendering of the ordinary Latin for "biggest rump," while the next most forbidding rendering is just an enrichment of the word English peasants in the Dark Ages used for "end," *buttuc,* whence our stilted term today, "buttocks."

7:02 Closed Door Interlude

Two minutes into the day, and what is done with the extraordinary ability to walk? Most likely a hasty retreat to the bathroom.

A microphone on the belly would detect the distinctive sloshing noise that is the reason. To the three-and-a-half-inch-wide bladder, slowly dripping down from the kidneys all through the night, has come the yellow-colored substance now eagerly to be voided.

There's usually a bit more than a pint — two large glasses' worth crashing about this inland sea in the morning. It's less for someone who turned the heating up high during the night, for in such temperature there can be up to a pound's more sweating than usual, and fluid going out of the skin pores will not be available to go out elsewhere. (Urine is 96 percent water — the other 4 percent a mix including urea, salt, sugar, proteins, fat, vitamins, and coloring from bile pigments. It contains two ounces of solids per day, and is about 1 percent heavier than water, which is why it slowly sinks.)

On the morning after a hearty drinking bout, by contrast, the delicate pressure sensors lining the bladder's storage space will be pressed on with more power than usual. It can be a greater impetus for getting out of a comfortable bed than tenuous thoughts of the day at the office to come.

The curious question is why this need leap into prominence only upon getting up? It could have roused the sleeping one hours earlier, but usually it doesn't. How come?

The reason is that the bladder is an uncommonly adaptable organ. When empty it shrivels to a small prune shape, thick and rugged to the touch. But when filled up from within it swells and swells stretching its cells sideways to fit the demand until at its full extent it is thinner than onionskin paper, just a few cells thick.

That taut stretched state can be reached as early as 3 or 4 A.M. When that happens, always ready sensors in the bladder wall send messages to the lower part of the spinal cord, and that in turn, without any

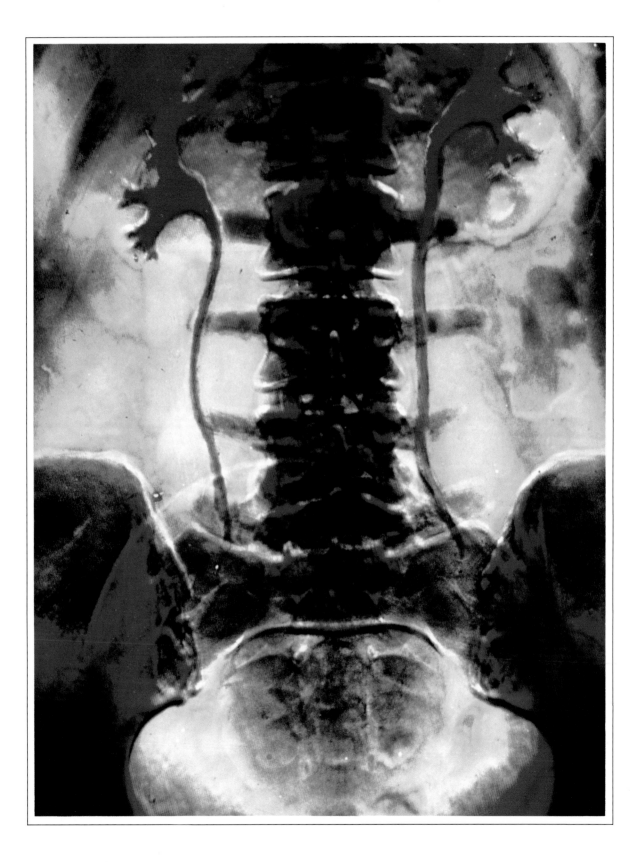

Normally the bladder wall is a tough muscular barrier (top), but at maximum diameter — as is likely to occur after a night's accumulation — the bladder's stretching wall can be just several cell layers thick (bottom right), which increases pressure on the stretch sensors just under the surface.

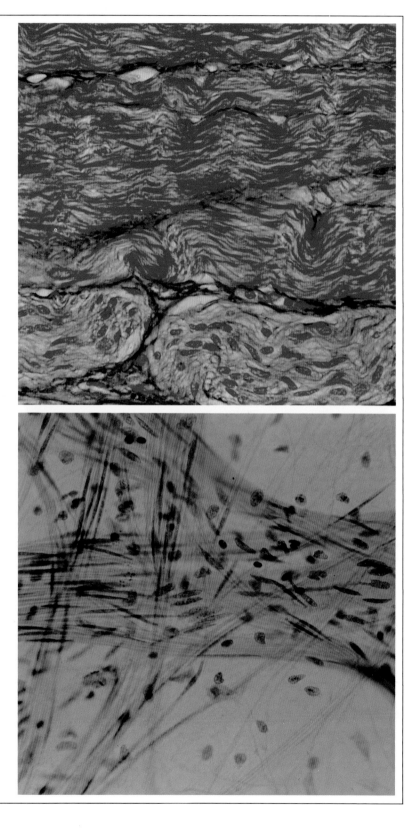

X ray with tracers highlighting path of urine from production in the two kidneys (red, upper left and right) down ureters — which pump the urine in rolling waves — to storage in the bladder (lower sphere). ◄

notification of the conscious parts of the brain, sends response signals with the power to trigger muscle back to the bladder wall. That contracts the bladder, and loosens the ring of muscle around the drainlike hole at its base, through which leads the path to relief of the pressure within.

Luckily there is another ring of muscle just beyond the one under the bladder's indirect control, and this second ring is held tight by centers solidly entrenched in the base of the brain. Those circuits lock when the brain that surrounds them has gone to sleep, and so automatically stay closed as long as sleep continues. (Older men are liable to be awakened in the middle of the night by a pressing need to urinate; that's usually because of the special case of an enlarging prostate gland that is pressing into the bladder, or that has narrowed the exit path of the bladder so much that it could not be fully emptied the evening before.)

When sleep ends, however, that brain center once again starts noticing the nerve signals running up from the bladder. Since those say ever so strongly that the storehouse is full to bursting and must be emptied, it's only by conscious commands to keep the second drain ring closed that a sudden spouting can be averted. The sound of running water or the sigh of gushing faucets can be especially distressing at this time.

All that's needed for the enclosing ring around the second outlet of the exit path to spring wide open is to stop this conscious intervention. Muscles around the bladder squeeze in rippling waves to help the torrent on its way, and the breath is briefly held. It's almost impossible to start urinating without holding the breath — as may be seen by trying. Only that way, with the glottis (the opening between the vocal cords) closed, can the sheet of muscle under the lungs be pressed down to further the sac's emptying. Bliss.

In the privacy of the home there's rarely a problem in this multicheckpoint process. But occasionally for men, similar efforts in their office building or at a restaurant later in the day will become trying events if there are others around. Any slight nervousness makes the abdominal muscles tense, including those clenching tight around the exit of the bladder. If the nervousness goes along with hastened breathing, as it often does, then there will be even less chance of that brief holding of the breath so needed to get things moving. Procrastination is of little help here, for always, and always harder, the bladder's call goes on.

7:10 Jogging

The graceless shamble from the bed in the morning can be transformed, a few minutes later on the sidewalk outside the house, into the much more fluid locomotion of running. Curiously enough, the early morning shuffling had an important purpose at its end, to wit the visit to the bathroom, while the apparently more purposive firmly stepped striding is not likely to be finished when a circuit of a few streets has been completed, and whatever was being chased has been caught, but is likely to go on and on in aimless circles for quite some other reason. What exactly this reason is had best not be examined, for the ramified depths of a devoted jogger's mind know no bounds, as will be affirmed by all who have shared a long car ride with a loquacious jogger in an evangelical mood.

But whatever the cause of the movement plenty of running-enthused physicians have done enough studies on exhausted runners, chasing after them with recording bags to breathe in, or sharpened needles to take samples with, to find out a lot about what is going on inside the muscles when someone runs. The doctors who run after runners like this tend to be unusually persistent individuals, and of their band one of the most persistent ever was a certain soft-spoken Oxford-trained researcher. After he had tested enough exhausted runners he got the idea that with the proper pacing technique there was no reason why a healthy man should not be able to beat what at the time seemed to be an unbreakable barrier, the four-minute mile.

While graduates of other institutions may have been content to write a paper on this and leave it at that, Oxford inspires its sons with tougher stuff. The ectomorphic doctor resolved to perform the experiment on himself, and stuck to his regimen with scientific certitude until, on a cool May 4, in 1954, on a track in the shadows of Oxford, he became the first human ever to run a mile in less than four minutes. Dr. (now Sir) Roger Gilbert Bannister had shown what Oxford men are made of.

Bannister based his run on the fact that no muscle can move unless there's some energy to get the interwoven strands it's made of to slide along each other. In all humans the energy suppliers needed to get the muscles going take the form of a delicately curving molecule, called ATP, which has something like three taut bowstrings stuck on it, all made up of phosphate like that found in certain brilliant parachute flares. If the third and final of these bowstrings shoots open and releases its energy at the right place in a muscle cell, then the muscle will move. If it doesn't, the muscle as well as the tendons, bones, body, and hopes attached to it will not get anywhere. So what's really going on inside a moving jogger is that these phosphate bowstrings in his ATP are continually being twanged, each time causing a bit of muscle to act.

This just puts the question one step farther back. What is it that charges up the bowstrings in the ATP, so allowing them to fling forth and get the muscles moving? Something must be providing the resupply, for the amount of already taut ATP in the body at any one time would be wholly used up in less than a second's movement. Only in the last few years has it become clear that there is not just one something that keeps the ATP charged up to move the muscles, but three distinct devices. Bannister's research, with others', helped reveal them.

The first is a chemical that is to be found liberally poured throughout the muscles. Called phosphocreatine, it's like a spare supply of already primed and taut bowstrings just lying around near the ATP in muscles, ready to pop into place whenever a gap opens in an ATP molecule that has just fired off its own taut bow in contracting a muscle.

It would be as if the English archers at the medieval Battle of Agincourt had stood in a field strewn with already primed bows, so that after firing the ones they had come with they could just reach down and pick up others to shoot.

This chemical is sprinkled all over, but there's only enough in any one muscle for about six to eight seconds of hard work. It's a useful amount to have on call, for there are a lot of brief actions, such as leaping onto a boulder away from a saber-toothed tiger, or sprinting for all you're worth the last few yards to the departing 8:15 train, that you need to be able to do without advance planning when the need arises.

For most competitive running events, however, this six to eight seconds of ever-ready oomph is not enough. Only sprinters in a 100-meter race are partial exception. Their distance is short enough so that they can get most of the way to the finish line before using up the phosphocreatine.

Most of the way, but not all Olympic sprinters break 20 mph in their dashes and burn the chemical at a 14.4 horsepower rate to do so. Once the phosphocreatine already in place in their muscles is used up though, usually around 80 or 90 meters down the track, the sprinters get exhausted. Nor are matters helped by the facts that three quarters of their ATP breakdown goes into useless heat, and that their awesomely pumping thighs are actually squeezing so tight as to block most of the ordinary blood flow through them. It's no wonder that most 100-meter races are won not by the runner who started best or went fastest in the middle, but by the one who slowed down the least before the end.

Because of this problem, sprint records have remained almost unchanged in this century, dropping only from 10.6 seconds for the 100 meters in 1912 to not even 9.0 seconds today. Other distances have seen far greater improvements. Sprinters would only join them if they could somehow finish races before all their phosphocreatine was used up, but there's little hope for that.

The second way ATP is charged to get your muscles working when you jog is through blood sugar. It comes into action when the first system peters out, after around 8 seconds along. The right sugar percolating over from the blood to some ATP that has used up its phosphorus bow can help shape the used molecules into a new energy bow, taut and ready for more action.

Back to our Agincourt example, it would be like a helpful sergeant running over to the archers who had used up their arrows and warped their bows, and showing them how to straighten the bows and quickly pick up used arrows. It's not as fast as the first method, which was to just resupply taut and strung bows, but it has the same result. ATP molecules recharged in this way are ready to go back to the leg muscles of

Muscle fiber held tight in a mass of the bodily connecting tissue, collagen; these holdings are crucial in keeping a runner's body intact.

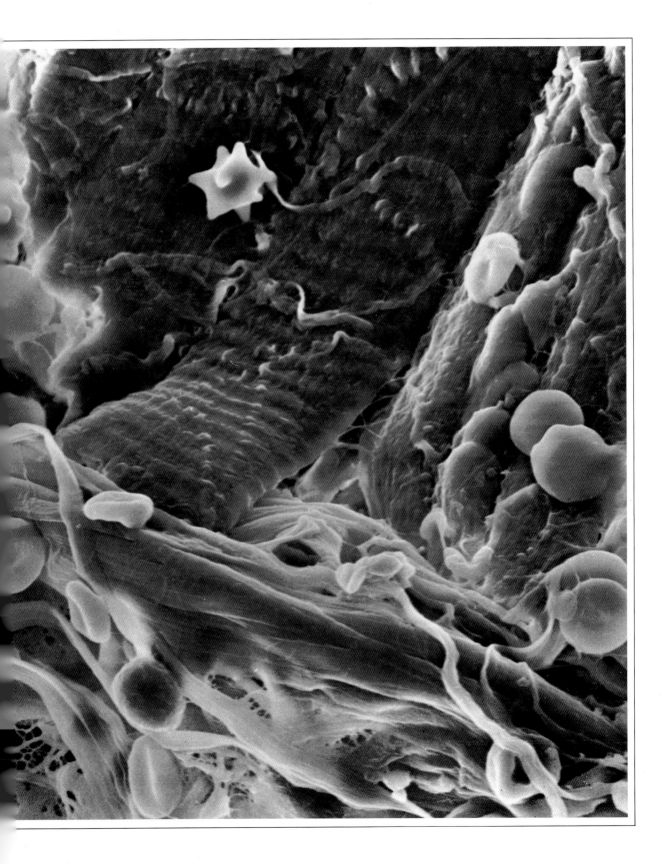

an eager jogger, and tense them again and again as is needed to keep on running.

Although blood sugar is what does this, eating straight sugar before hard exercise will not have the desired effect. Linda Fratiane, America's former Olympic figure skater at Lake Placid, reportedly took several mouthfuls of honey before the national championships in 1980 and proceeded to go out on the ice and fall three times. The body has better uses for sugar than to have it lying in the muscles, on the off chance that some crucial use will sometime be made of it.

Most of the sugar that we take in is instead converted into "glycogen" — which is more or less how an ancient Greek would say "sugar producer." And that's what it is.

When sugar is turned into glycogen it can't be used by the ATP in the muscles but is instead neatly put into storage in the liver. Only the proper hormone, inspired by the start of running or other exercise, will shake some excess glycogen out of the liver and send it to the muscles, there to be broken down into the right kind of blood sugar (glucose), which can go into recharging the crucial ATP.

This second process picks up after the six or eight seconds when the first one leaves off. It hinges on the right hormone's trickling down from the brain, and through a domino chain of events jarring the stored glycogen into a place by the ATP where it can do some good. But how do we get the hormone going?

A sure way is to start jogging, so using up enough blood sugar to send this inciter of more on its way. That's perhaps the best reason for loosening up before a race. Few athletes are going to pull a muscle in a run they may have done a thousand times if they don't loosen up, despite what the coach might say. Rather, their loosening up gets the right hormones for releasing sugar on their way sooner. Perhaps for the same reasons, ordinary joggers are likely to have a better session on the track or in the park if they've thought about it for a few minutes before, rather than starting abruptly before the brain has any inkling of what glycogen goodies it might have to supply. (As an added attraction, when thinking about exercise, the body often automatically squeezes a pint or two of extra blood out of the liver in preparation. The left ventricle of the heart is likely to quiver and move more powerfully — an important exercise preparation too.)

This second process too has its limits. After about two minutes a sludgy by-product of its chemical juggling acts begins to be produced in too much quantity. That by-product is lactic acid, a chemical quite close to some ingredients of ordinary milk and thickened yogurt. It's fermented out of glucose in the muscles in something of the same way alcohol ferments out of bacteria in the brewery. This lactic acid stays in the muscles, next to the ATP, which is charging and unleashing its energy, and acts like an increasingly thick sludge that gets in the way and stops the ATP from recharging. It's what produces the common feeling of a sudden seizing up and tiredness. More blood flow might help, but the body couldn't take it. Alaskan sled dogs who have been measured doing 70 consecutive four-minute miles do so with a pulse rate over 300. No human jogger or even Olympic athlete will get near that!

Lactic acid buildup often goes along with the increased panting caused by the heart's trying to keep the muscles working despite the mucky sludge. The result is the unpleasant exhaustion that sends so many earnest first-time joggers off the track before they've even gotten a chance to start one of those long zen-inducing runs their friends have told them about.

Their friends tell them this because they, those lucky, hardy creatures, get their running energy from the third process. It's the best. In it fresh oxygen pours into the tiring muscles, and so allows the blood sugar to recharge the ATP without producing any of the unfortunate sludge. There's no lactic acid when a little oxygen gets in there. With the oxygen pouring in, other things than glucose can be used to stoke the ATP in the muscles, including carbohydrates, fats, and proteins.

The transition to this third process is probably what is commonly called getting your second wind. To those who don't get it, it must seem a mystery, but to those who do, it is what makes long-distance jogging the eminently pleasurable experience it is.

That's because the great amount of oxygen flowing to the muscles is not just shunted there and nowhere else. It's also in the blood that goes to the brain, there seemingly to help great thoughts flash along more quickly and clearly than they do when only a paltry supply of oxygen is getting up to the seat of learning.

This joyously running body is undergoing even more contortions than it did when merely bouncing, shaking, and squelching in its low-key walk across the bedroom floor. Full-fledged jogging surpasses this in every compartment. The jouncing eyeballs jounce even higher, squishing flatter than a sphere as they splat into the top of the eye socket with each downward step. The cheeks flap even wider, and the intestines smack up against the liver with almost a wet squelching noise. (These effects, of course, occur only on quite a minute scale.)

Bones in the fingers and wrist are tugged so hard by the centrifugal force in a jogger's arm swinging that they would break loose from the arm and tumble into the pouches formed by the skin at the front of the fingers, like so many bolts being heaved into an empty glove, if they weren't strapped in place by the continually flexing muscles that stretch down to them from the lower arm.

Along with the tugging is the turning, as the legs twist around the centerline of the body, winging in from the outside to get the best stepping position straight in front, and the arms above them whip around in the opposite direction to keep the turn from ending like a figure skater's final spin.

As the legs go, so do the hips. Just as the gyroscopic tendencies of the legs had to be countered by an opposite twist from the arms, so the tight twist of the hips is countered by an opposite twirl from the shoulders. What this does to the top of the arm is terrible to behold, for the arm starts on its pendulous arc a moment before the shoulder does, and in that moment it is pulling away from the shoulder with a wrenching jerk strong enough, if there were no help, to pull the top of the armbone clean out of its curved socket in the shoulder.

But there is help, in the form of the ligaments, dozens of tiny braided fibrous strands, white and strong, which have one end tucked into the shoulderbone, and the other pinioned down to the top of the arm. When the arm swings with a jogging step, the strands snap taut and hold steady, keeping it next to the shoulder until that upper support begins to turn too.

The arms aren't the only limbs to be precariously jostled in jogging. As one muscled leg goes arcing forward, pumping in the air between strides, the flesh on the thigh starts a truly grotesque squidgy roll from the inside of the leg to the outside. On someone with fat legs it's a veritable tidal wave of jiggling flesh. In fat and skinny people alike the long thighbone tucked away in the center of this heaving gets in on the roll and almost as if being cranked by tiny men with pulleys starts twisting outward too. Mercifully, the motion lasts only until the churning leg touches down on the ground.

Just before that auspicious contact, with the foot still a quarter-inch above the path and hurtling down at top speed, the tiny bone in the heel acts as if it realizes that a most uncomfortable fate is in store for it. For the heel bone is the one that will hit first in slow jogging, and it hits with the body's whole weight coming in on top. To spread the damage, associated tendons and muscles quickly loosen their supports and hang suspended in the back of the foot as innocuously and limply as they can.

The moment the foot lands the bones get a splaying squish, but it's not to be for long. In about one third the time it takes to blink, the rest of the foot in even a plodding jogger goes rolling off in front. Loosened up as it is in that brief time, the heelbone can take the pounding.

But the heelbone is not the only one in danger. All the bones in the front of the foot are now faced with the same slamming impact in turn, and just as the heelbone did, they all sensibly loosen up their ties with each other and try to hunker down in the soft tissue underneath when the behemoth that is their owner passes overhead. It's no wonder that numerous species, humans too, put some adipose (fat) tissues on the soles of their feet to help take up the shock.

For one brief moment the foot is bulged outward, splayed like some pastry dough being slammed on with a rolling pin. But the squashing is brief, for the inflicting body has so much momentum that it's only too quick to continue on its jogging way. That's when the bones in the foot and ankle get their own back. Squashed out flat under the full body, they could only cower and hang loose. But once it is no longer directly above, and the leg is ready for takeoff, the bones down there snap together into the tightest formation they can achieve, providing an instant rodlike lever where what a moment before had been a jellylike foot. In their new straight shape they propel the body forward and away.

It's as if an unassuming twig caught under a boulder suddenly transformed itself into a crowbar and propelled the stone off. The stomping jogger gets a powerful kickoff from the suddenly tautened footbones, a push which seems destined to end all the problems down there. For one incredible moment he's going forward by having his power leg planted and moving backward away from the body. But alas, in just a moment, and after a bit more thigh blubber bouncing, the whole trodding process is to get repeated again — perhaps two thousand times in a mile run.

All that's wearing on the feet, but the good feeling a jogger gets for his labors is not so easily renounced. A lot of the pleasure comes from the oxygen high of steady exertion, but some of the good feeling comes about because of more obscure bodily changes. As we shall see in a later chapter, around most contracting muscles and moving joints are special senses that send to the brain electrical impulses that change with each movement. The electrical patterns produced this way in the brain during running will never happen in any other movement, and it's possible that they fit in with and enhance deeper rhythms that are inherent in the nervous system. Less speculatively, another important change in jogging takes place in the bones, a part of the body that is as much a living organ as any other.

Bones are usually thought of as white and sterile, but that's just because skeletons put on display have been carefully boiled and cleaned first. (The very word *skeleton* testifies to this process, for it is the Greek for "dried up.") In a healthy panting jogger the bones are pulsing and soaked with red blood. This is useful on top of being colorful. The bones are plugged into the whole blood circulation because they have a lot to offer it. Deep in the center of some of the strongest bones — the skull, backbones, and ribs — is where white blood cells are made: over ten million come out a minute, on the average. That vast number is needed to course around in the bloodstream and gang up on any dangerous germs that have somehow gotten in where they shouldn't.

It sounds as if the poor germs would always be outnumbered, but every time you catch even a cold it's the body's defenders that have been swamped. The rumbles, crunches, and shakes a jogger produces in his body seem to squeeze on the compartments inside the bones where white blood cells are formed, and indirectly send even more than usual of these disease-resisting wonders popping out into the bloodstream. And while the fewer colds that runners report can't be due to the white cells (which do not affect colds or other viruses), they can be due to more antibody or immunoglobulin proteins floating amidst the inorganic salts and other substances in the wet blood plasma. The several quarts of lymph fluid inside you, that carry so many defensive antibodies and white blood cells, can be pumped by heaving leg muscles up to ten times more quickly than when at rest. Altogether it gives a feeling of smugness and good health.

Jogging's hard bouncing also brings out endogenous pyrogen, a normally reticent small protein you have that defends against illness by such clever stratagems as pulling out of the bloodstream some of the iron that bacteria need to feed on and secreting it safely out of reach in the liver for the duration.

Increased running also seems to stimulate growth hormone from the pituitary gland which lies below the brain. This is the hormone that gets squirted out in carefully regulated amounts until the early twenties to make bone-lengthening cells grow. After that it helps close scratches, build new bone when it is needed, and generally help tissues grow when they've been worn down. How running raises its level is less clear than how bone jostling releases white cells. But it does seem to be just as frequent, and just as beneficial.

Another pleasing effect is that a few months of hard jogging can get the body to carry its fats through the bloodstream in tightly concentrated little spheres instead of in the usual loosely splayed wider ones. The tighter spheres have less space in them for cholesterol (boo, hiss), and so what amounts of it they do carry are tightly packed in, and not too easily sloughed off. The blob-shaped fat spheres of more lethargic people do not seem to have their cholesterol so tightly attached, and so are more likely to let it leach off and find its way to coronary arteries and other unwanted places.

These changes happen equally in both sexes. Others do not. Many female joggers, for example, find that they stop having monthly periods. Although there can be many reasons for that, it's often because they've burned a lot of the fat that they'd had in their bodies, trimming the figure from an ordinary American's 40 percent or more to a trim runner's 12 to 18 percent. Now it's extraordinary that adipose tissue stored in subcutaneous fat on the hips, face, or arms can affect hormonal influences on the inner lining of the uterus, but it does. Inputs to the pituitary gland do not register enough fat stored in the body to support a pregnancy, and so the gland cuts back on the hormone schedule,that produces the regular periods. (That's one reason why women in India don't start menstruating until the age of sixteen or so, an average four years later than their chubbier American counterparts. It's rarely a reason to stop running, although numerous worried gynecologists, invariably male, and invariably nonrunners, might at times recommend doing so.)

7:45 Eating

After the labor, the reward. Some nice, wholesome food is a natural recompense for the virtuous tiredness of a morning's jog. It is a wholesomeness, however, that is mercilessly challenged from the moment it enters the mouth.

In most Americans' mouths can be found an extraordinary range of bacteria, viruses, and fungal growths, including such delightful-sounding ones as viridans streptococci, corynebacteria, bacteroides, fusobacteria, diplococci pneumoniae, hemphilus influenzae; sometimes streptococcus pyogenes and neisseria meningitidis too. There are more, lots more, and the total population is well above the one hundred million level.

There's no way out. Even a thorough brushing of the teeth half an hour earlier would have done little to stanch their numbers. Indeed one of the main sources of cavities for children are bacteria that have propagated on their toothbrush!

These intruders are too small to show up in a casual glance by any companion sitting on the other side of the breakfast table, so few people are likely to have their lack of mouthly solitude brought to their attention. That is lucky, for it would possibly lead to only the most

A morning's mouth. Above, bacteria on a freshly brushed tooth; below, closeup on a clean tongue, showing mucilage, saliva, sloughed-off cells and pitted taste-detecting papillae mounds.

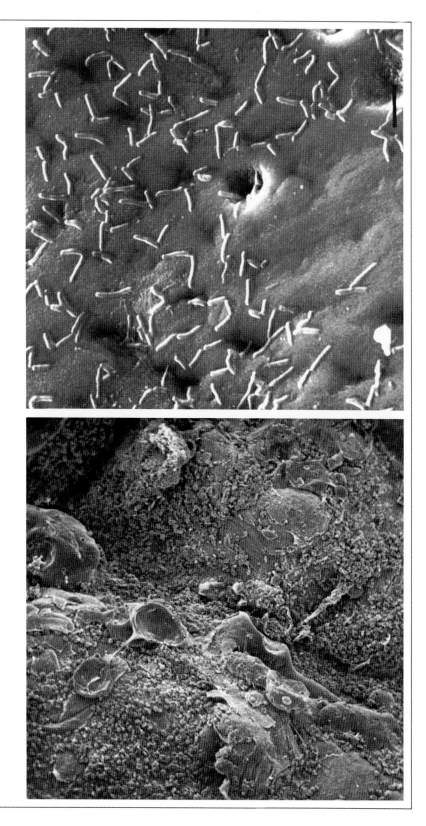

unpleasant of reactions, perhaps including a refusal to let any food slide through the contaminated orifice ever again. ("Sorry to bother you, Susan, but you seem to have a few visitors." "Where?" "In your mouth — AAARGH!")

Though understandable, that would be excessive. The millions of microscopic residents in the mouth are for the most part totally harmless, living peacefully amid the saliva and the teeth, and producing damage only when they propagate to levels far above the normal ones. Any moist tissue of the right acidity will do for bacteria: the vagina has an incredible number of active bacteria too. Mouth bacteria live in conjunction with a whole interlinked zoology of minute creatures all over the face, and since each group has its allotted niche, no one of them is likely to find the necessary space to breed to harmful levels.

The niches can be precise indeed. One crucial element of this shared economy, found in perhaps 10 percent of all Americans and Europeans, irrespective of health or social class, is a little mite that lives in the spaces around the base of the eyelashes.

Wedged in there headfirst, it's shaped like a crocodile with eight legs and a wriggly body; the males have the noteworthy distinction of having two penises, both protruding rigid from their backs, while the females have two corresponding orifices set high on their chests. The mites' presence keeps the eyelash bases from serving as home to any other inhabitants wandering from the mouth, and so reduces the chance of their growth to dangerous levels.

These proudly individual hordes exist, reproduce, and scurry about in our mouths and on our faces without producing any awareness of them. That's why the first act we usually think about in eating is the eager chomping of the teeth themselves. This, too, is more involved than we think. Even with the most delicate bite down on a piece of lightly toasted bread, the teeth that do the biting twang like a plucked guitar string, as their outer enamel shell charges down onto the softer tissue pulp within, only to bounce up again as the pressure of the bite whiplashes up from the jawbone on which the tooth is based. During serious chomping this twanging can go on in all thirty-two teeth at once.

In children the tooth noises are even more arresting, for they can have a whopping total of fifty-two teeth, most resting patiently in the gums below the aptly named children's teeth, and all fifty-two will twang and resonate with every eager bite. It's a veritable dental symphony, lost to most of us with our imperfect hearing, but perfectly recordable with properly placed microphones.

Early investigators didn't have the equipment to notice these sound effects, although to be fair, many of them took only a passing interest in the untold mysteries of teeth. The normally diligent Aristotle, for example, wrote that women had fewer teeth than men, a fault he could have rectified by asking Mrs. Aristotle to open up and let him count.

Chewing is useful only if there's something in the mouth. This fact is often taken for granted, but bringing the food there, and getting a hold of it in the first place, is a complex maneuver. Time-consuming too: in an average twelve-minute breakfast many people will spend a sixth of their time just moving their arms back and forth, as in some sort of bizarre shadowboxing ordeal, transporting their nutritional essentials upward in some thirty lifts of four seconds each.

Just reaching out to grip a glass of orange juice rich in vitamin C takes much accumulated practice, as the joyous mess-making of an infant so exuberantly demonstrates. A five-month-old child takes a parentally exasperating 16.5 seconds on average, to reach for something on a table in front of him, and does so with an armlocked sweeping of its surface.

At seven or eight months his trigonometry is slightly better, and most toddlers will reach down in a halting arc, often to overshoot the target, but still safely clearing intermediate obstacles such as plates or already spooned out food before them. By eleven months their reach will be straight and direct, and with only a brief pause of hesitation before making contact with a desired fruit juice glass will clock in at a much improved 2.3 seconds.

Adults get it under 2.0 seconds, just, by scooping up the glass with nary a trace of a pause, a figure remarkably constant from the most refined of dinner parties to a furtive between-meals grab.

Once your hand has reached its assiduously sought after prize, this glass of freshly squeezed orange juice (as your body is some 60 percent water, drinks are invariably reached for first at breakfast as a precaution against drying out), there remains the need to get hold of it. Nothing is as effective for this as to wrap the thumb around one side, and the fingers around the other. But the act is not innate.

A five-month-old who's made it to his juice cup will take an average 11.5 seconds, first to press its little finger around as much of one side as it can grasp, then to send in the rest of the fingers in a clump, and finally to squeeze firmly with the base of the palm to bring up the rear. Even then the hold is a precarious one, though infants' great saliva bubbles of delight upon achieving it are likely to obscure the fact.

As always, insouciance is the price of speed, and the adult who can form a good grasp on his breakfast glass in what is on average a third of a second is likely to forgo all pride in the act. But he shouldn't, for an amazing bit of mechanics has just been performed. It is accomplished with that rarely applauded remnant of our reptilian past, those objects that were once vicious scales, and are now delicately coiffured: the nails.

Without nails as an immovable support, the skin of the fingertips would slip over and around the thin bones within the finger. The result of such slipping would be to start the breakfast glass spinning merrily on its axis as the hand closed on it and clumsily spun off again; an admirable, but not particularly efficacious, feat.

Nails are crucial in another way too. They serve as a base to support the fleshy parts of the finger that do the actual grasping. These fleshy parts are a lot wider than the bones under them. That's what gives them the area needed to build up enough friction to keep the lifted glass from sliding loose and plummeting down to the table.

The fleshy fingertips get even more friction from undulating ridges

Little boy having one final peer before reaching for food; a pleasing yet demanding task.

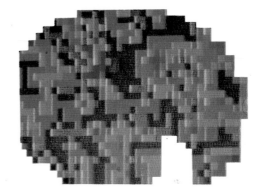

Above, a real-time computerized image of metabolic products in the brain of a quiet person thinking about a task; below, the precisely localized brain activity that must be produced to be able to grasp an object.

of their surfaces, ridges made of the same substance that gives the human tongue its uneven feel, and the cat's tongue its definitely raspy one. They are what produces our fingerprints.

Like a radial tire gripping a slippery road, fingertip ridges spread wide to provide even more friction for the holding. On top of them (or under them really), the whole fleshy pads on which the fingertips rest squish about in just the right fashion — in a bit here, out around the edges there — to get the tightest seal they can.

Most of these unthinking grasps for the morning's sustenance are likely to come swooping in from the right. They are likely to be less unthinking if they come in from the left. For the world's left-handed minority has often been reminded of their obscure and revolting manner of reaching for things. In French speaking countries the word for "left" also means "uncouth," and in Germany it is the base of the word "awkward," and in Spain and Italy it's little better. Attempted cures for left-handedness have included tying a child's wayward hand to his belt during school hours so he couldn't write with it, as was done in certain parts of the United States until recently, or even immersing the offending appendage in a hole filled with boiling water, a charming custom reported by one visitor to a West African village early in this century. Attempts at discouragement since then have become disguised, but never relinquished.

Something like half of all seven-year-olds prefer their left hand in some ordinary actions, but three quarters of these are cajoled, induced or forced into becoming exclusive right-handers by the time they finish puberty. That leaves a bit over 7 percent of the population as confirmed left-handers: fifteen million in the United States. All of them are liable at one time or another to reflect on the meanness of their fellow creatures, when searching for nonexistent left-handed scissors, golf clubs, potato peelers and, as the most ignoble of discriminations, bathrooms designed with toilet paper holders on the left side of the bowl. Some have retaliated. Elias Howe, left-handed inventor of the sewing machine, placed the needle on the side where he could get to it best, and patented the result. All right-handed sewers who have caught their undeft left fingers under the needle of a sewing machine owe a debt to Mr. Howe for being thus allowed to repent for their sins.

Left-handed parents have a better chance than most for some solidarity at the breakfast table. Half the children of two left-handed parents are similarly inclined in turn, while even in marriages of divergent directionals about three times as many of the kids will be left-handed as in the general population. The mighty Kerr clan of Scotland were specialists in leftistic inheritance. Many of their castles had bastions with outer steps that spiraled up counterclockwise, so that a left-handed swordsman could back up the turret and still hew away at the enemy with the greatest of ease. (Normal castles have their stairwells rising clockwise for ease of right-handed defensive hewing.) It was a wise investment, and successful defenses must have been plenty, for even today the odds are three times higher than usual that the next Kerr (or Karr or Carr) you meet will be left-handed.

All this action goes into just reaching for something at the table and getting a grasp on it. What's needed next is to bring the grasped object to where it's expected. That too is harder than it seems, for a glass brought straight up will be in an undrinkable position. (Try it). It must be tilted before reaching the top of its voyage, just enough for its contents to flow into the waiting mouth.

Frogs cannot tilt their forearms and are forced to make do with long and sticky tongues. We are luckier, for no one thinks twice about that crucial twist when bringing up the goods. Well one might.

Without it we would not only have to drink in a singularly wasteful fashion (tossing the desired fluid straight up, and trying to catch as much as possible with a forward lurch of a stretched-open mouth, with all hostesses of good breeding gracefully ignoring the portion that missed), but without twisting the forearm we would also be unable to turn a screwdriver, strike a flint, carry a tray, lift a phone receiver and turn it to the ear, play tennis, or scratch the back of the head.

What helps is the fact that our forearms are made up of two long bones, not one, and when they turn in a slide over one another the whole forearm turns, too. For this power we are indebted to the crossopterygian, a creature that, once thought to be extinct these past 300 million years, was the direct ancestor of ours that first used its fins as limbs. In the foremost of its flippers it had two bones, lying side by side and so readily twistable; the same bones, by direct inheritance, that have developed in us to serve in the most indispensable of food-gathering techniques: the stretching of an arm over the table.

As our contented runner sits there chewing breakfast, his salivary glands are well in on the action, first swelling and then extruding in a gurgly squirt their sticky contents.

The squirting is carried out through relatively long tubes. They're needed because the delicate saliva producers would be bruised if they were on the surface of the teeth or cheeks, right at the scene of the chewing. The most distant of the three saliva glands in fact is placed way back beside the ear, and emerges into the mouth from little funnels beneath the rearmost molars. The gland is usually safe back there, with its only trouble being the occasional overswelling called mumps, which is due to an excessively wandering and infectious virus.

From the normal saliva glands come up to two pounds of a mostly watery fluid each day. Floating in there are several food enzymes, including the substance called amylase, that is incredibly good at tearing apart carbohydrates into sugars, and is what's working away to make ordinary bread taste sweet as you chew it. Also in the daily two pounds are a whole host of motley additionals, including bicarbonate buffers that wash the entire mouth cavity, antibacterial agents, and a potent mucus-producing protein that makes the saliva stickier and a better lubricant — an effect so important that exactly the same protein is produced in the vagina and secreted during sexual arousal.

Saliva also comes into the mouth containing some of the waste products from the glands where it was produced, including urea, that yellowish substance last found in the cerebrospinal fluid, and usually found sequestered away down in the bladder. For those to whom this regular addition to the saliva sounds distasteful, it should be pointed out that the slight steady dribbling of ureas is probably preferable to daylong accumulation in the saliva glands, with the resulting concentrated mouthful that would then be produced. Even more importantly the enzyme amylase comes squirting out in the saliva. It's what unstitches carbohydrates into sugars, and so makes ordinary bread taste sweet as you chew it.

These are necessary actions but not really graceful ones. Not so for the tongue. As master of ceremonies in the morning's chomping, the dexterously weaving tongue can push a bit of food back between the teeth, while simultaneously rolling an already chewed bit across its middle for a good saliva dunking.

This dexterity is recognized by the fact that foreign languages are commonly referred to as foreign tongues; "language" itself coming from the Old French and Latin for tongue — witness *langue de boeuf à l'estragon*. The tongue's usual agility is also recognized by the sensation of near total surprise we feel when it fails for an instant in its ballet, and gets momentarily caught between the teeth for a slicing chomp. The right side of the tongue is more agile in right-handed people, while the front one and a half inches are the most graceful in all, having extra nerve controls and playing a key role in most languages.

(Dogs by the way are even better endowed in the tongue department, and manage the apparently impossible task of lapping milk straight up by curling the tip of their tongue backward and around, so producing a rear-facing ladle with which to scoop up their beverage.)

The cheek muscles, "buccinators" in official nomenclature, are also crucial in holding the morning's food balanced properly on the lower teeth to be chewed, as attempts to chew while holding the cheeks totally loose will demonstrate. Here again the name is a giveaway — *buccinator* is the Latin for a trumpeter — an individual who needed stiff cheeks to get a good sound out of his instrument.

Only when food has been chewed can swallowing usually be indulged in. This multistage technique begins with the tongue lifting the suitably prepared lump of food to the top of the mouth, thence squeezing back as far as it can reach. A food ball so placed is in a precarious position, for suspended under it in the back of the throat are the two well-known tubes, one to the lungs and one to the stomach, and into only the latter should the food descend.

This crossroads problem is produced by some unfortunate engineering we all suffer from. The food intake slot starts out in front of the air intake (mouth in front of nasal passages where they head down), but then switches for the food channel to run *behind* the air channel (as indeed the stomach is behind the lungs). The crossover has to happen at some point, and the top of the throat is it.

To try to keep things in their proper conduit, several muscles are positioned to provide a temporary right-of-way for the food about to be swallowed, like the temporary barriers put up at cross streets during a parade.

One of these barriers is the epiglottis, which is usually found resting

Bubbles of saliva — translucent, mildly alkaline, loaded with enzymes, and filling the mouth in abundance.

firmly above the top of the breathing tube. When a food ball approaches, the epiglottis will clamp tight over the path to the lungs, so that the food doesn't rumble down there by mistake. The chewed and thoroughly wet food flows around your epiglottis in two streams, rejoining just below it to continue safely toward the stomach. A trickle of food often slips through the epiglottis's seal, but doesn't get very far down the channel to the lungs because of a backup clamping by two folds just over the vocal cords. In extreme cases this clamping is audible as a choke.

All this barrier switching lifts the normally stationary cartilage surrounding the nearby voice box, and that lifting in turn is what causes the spectacle of a bobbing Adam's apple, an event that can embarrass a self-conscious swallower.

Women are spared being at the center of such a spectacle — thus the name — not because they swallow any less vigorously than men, but rather because their shorter vocal cords can be housed in a smaller and less noticeable voice box, the whole hidden beneath a greater fat layer under the skin. As a sign of swallowing's vigor in both men and women, though the very back of the roof of the mouth, the soft palate, actually gets sucked down the back of the throat to a point barely above the level of the bottom of the chin, and is let up again only when the gulp is completed.

Another barrier muscle to keep the air tract clear of swallowed food is the wriggly pinkish object we can see dangling down from the roof of the mouth when peering into a mirror and saying "aaahh" — the last clear sight before the dimness produced where the back of the mouth tunnel starts its curve down. Called the *uvula,* Latin for "little grape," which to those of a poetic bent it might be, it swings up when swallowing is going on, so making sure nothing untoward goes up to the nose.

Since laughing will keep the uvula from curving up tight, any swallowing of fluid that goes on when one is overcome with mirth is likely to lead to a most unwanted rechanneling of the swallowed substance, up to the nasal cavities and then out the nose. Soup seems especially conducive to this mishap, though the humorous nature of the fluid continues to be a mystery to science.

Once swallowed, the food goes down the rest of the way automatically. It doesn't fall, but is pulled through the esophagus (a particularly fearsome name this, but just Greek for "to carry what is eaten") by pulsing muscles that transport it the 10 inches to the stomach in 9 to 13 seconds, a rate that would convey it the length of a 100-yard football field in an hour and a quarter that means that a second bite is likely to be started before the first has reached its destination. Birds lack this feature of automatic contractions going all the way to the stomach, so producing the anomaly that while a man can eat while hanging by his knees from a trapeze, if so inclined, birds must lift their head for each swallow when drinking.

With this early morning activity completed, the rest of the day is ready to begin. Having already successfully handled brain-case decompression, spinal imbalance, leg muscle deenergizing and epiglottis

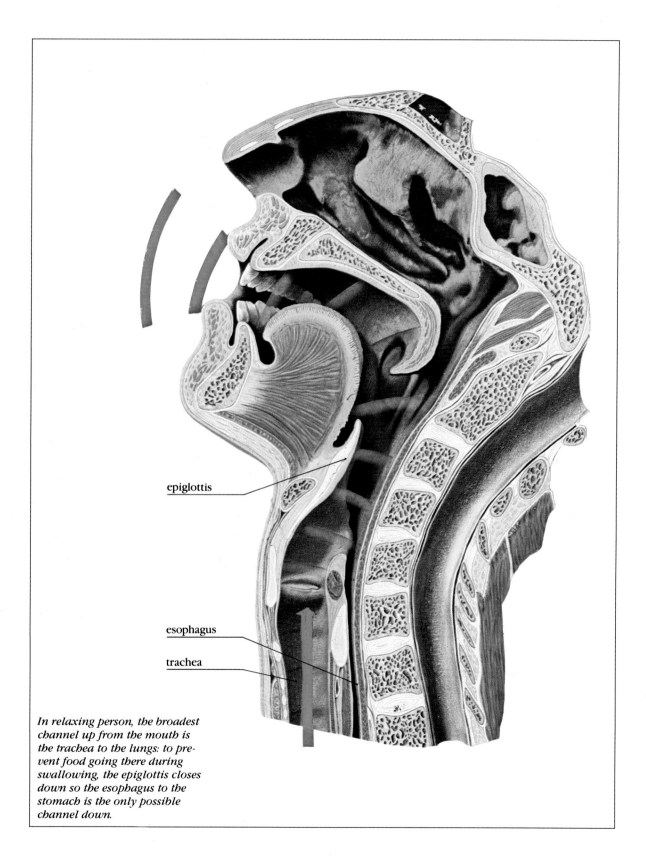

epiglottis

esophagus

trachea

In relaxing person, the broadest channel up from the mouth is the trachea to the lungs: to prevent food going there during swallowing, the epiglottis closes down so the esophagus to the stomach is the only possible channel down.

positioning, our greatest trials should prove no more taxing.

The key difference is that what has happened so far depended on the body's strictly mechanical, or structural, aspects, and could take place without the slightest thought. That is not the case for the more complex activities of the day, especially those dealing with the emotions, when our awareness of living within a body comes only too noticeably to the fore.

As we now shall see.

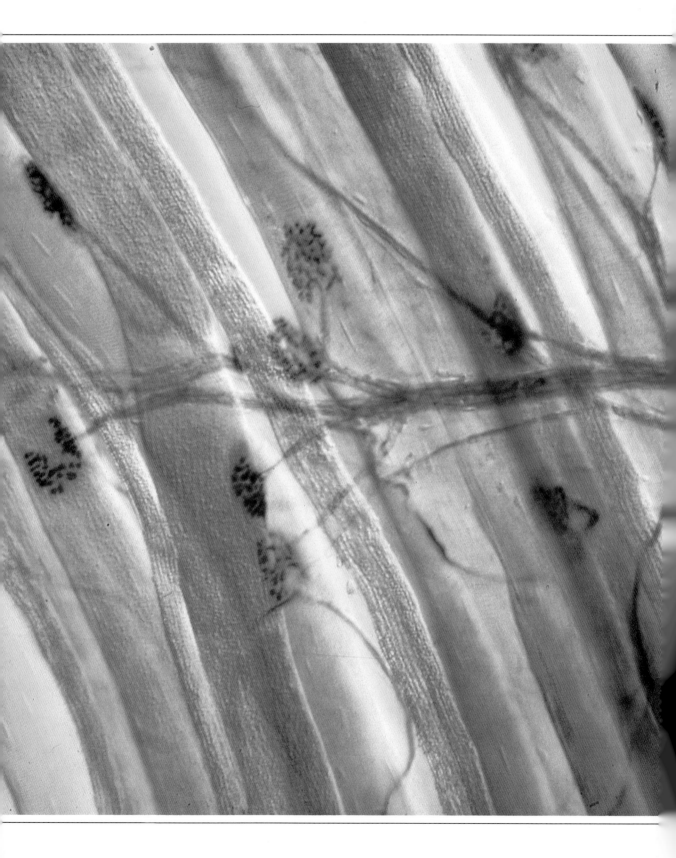

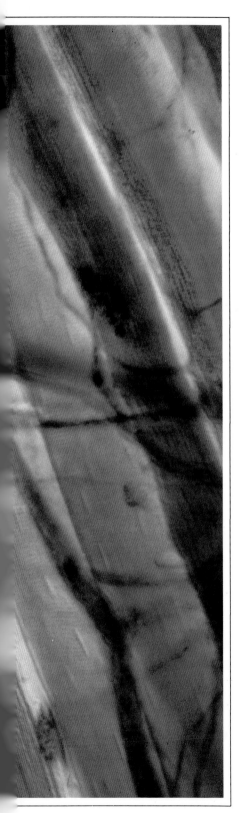

2
Emotions: Fear and Anger

Innate Aggression

First let's get it straight. Is man innately aggressive or isn't he? When you get mad at work, or lose your cool at home, are you showing a temper everyone has bubbling just under the surface inside them? Or are you giving in to more personal, less justifiable impulses? Scientists have been at work on this for several decades now, and they've managed to come up with a lot of suggestive evidence. It might be rough for those who like to think gentle thoughts about their fellow men, but on the grounds that "a liberal becomes a conservative once he's been mugged," it's evidence that everyone should have to face, whatever their presuppositions might be. Here are the key points that have been publicized so far:

1. Alongside the fossilized skulls of the proto-man australopithecene, who lived one million years ago, archeologists have found broken-off antelope leg bones that have two bumps on them precisely matching two cavities that are found on most of the skulls. The conclusion that early man engaged in bashing his fellows to death seems inescapable.
2. Even in modern times, primitive tribes whose members can express their natural impulses, tribes like the Apache of the Great Plains, or the ferocious Maori of the New Zealand lowlands, have spontaneously attacked and tried to kill all the peaceful settlers they could come across.
3. With the latest biochemical techniques it has been found that most of the dangerous psychopaths in high-security prisons, people like Richard Speck, the murderer of eight student nurses in Chicago, and

others, have been born with an extra Y chromosome. The sufferers are generally tall, strong, and unpleasant men, and their genetic imbalance shows that their violence is innately produced.
4. No less an authority than Konrad Lorenz, winner of the Nobel Prize for his animal behavior studies, has asserted that animals in the wild spontaneously attack each other with fearsome aggression, and that this strong inheritance must be one we all still have.
5. Finally, a Yale professor has been able to stop a madly charging bull by stimulating an electric implant in the aggression center of its brain, so showing how closely all mammals are under the automatic control of aggression.

That's the basis people who argue for the innate aggressiveness of man rest their case on. These facts are repeated over and over: in the writings of Robert Ardrey, Konrad Lorenz, and Bruno Bettelheim, in other books, informed TV talk shows, and even in-depth newspaper and magazine pieces. It seems a strong case, this story picture of humankind as a species filled with aggressive urges that are only poorly held back by the constraints of society. There is only one problem with this picture. It is false. One hundred percent, all encompassingly, false. Let's look at the points once again.

1. More careful researchers have gone back to the australopithecene skulls. The cavities in them do not precisely match the shape of the nearby antelope bones. What they do match is the outline of natural boulders that accidentally came to rub against the skulls in the hundreds of thousands of years they lay squeezed in the rock layers, getting fossilized in the first place. Not at all a sign of ape-men bashing each other; the researchers who said it was fudged their data.
2. The Apaches were originally a hunting-gathering people who were scared of horses. They got most of their food through their women picking vegetables and fruits, and lived peacefully in the forested areas east of the Mississippi. It wasn't idyllic, as they probably had gum problems and a high infant mortality rate, but it wasn't so bad either, and it certainly wasn't warlike. It only changed when European settlers got wind of them. Protestant missionaries gave them alcohol, VD, and kidnapped lots of their children. U.S. army forces pushed them from their usual, green gathering lands, to the arid semideserts of the American Southwest by repeated attacks with heavy rifles and field artillery pieces. Only occasionally did the Apaches try to struggle back, in a sporadic and isolated manner. Reports of continued and vicious Apache attacks were made up and fed by U.S. army forces to newspaper reporters for wide publicity. The truth of their information can not be expected to be greater than were the communiqués U.S. generals released a century later during the Vietnam War. Nor can the credulity of the reporters in the last century in copying down army statements and filing them with their editors be expected to have been any less than their modern counterparts either. Ditto for the Maori, and other so-called savage tribes.
3. As to the extra Y chromosome, the original study was done on one small ward of one Scottish prison in 1965. Since then other research-

Nerve endings. (overleaf)

ers have found that 99 percent of dangerous prisoners do not have an extra chromosome. About 100,000 male Americans do, and they are mostly to be found, as calm and nonlawbreaking as can be, in totally ordinary jobs such as computer programming, law, factory work, the priesthood and rabbinate, or medicine.

4. Lorenz might have a Nobel Prize, but this famous assertion on the spontaneous aggressivity of animals is one that he has had to try to waffle his way out of. The strongest example he can point to is the occasional twitching of one fin of the female oysterfish, in certain territorial and dominance displays, an action perhaps of interest to male oysterfish, but not one especially feared by it, nor one of obvious significance for the habits of mankind.

5. The Yale professor José Delgado made the front page of *The New York Times* with his account of how he forced a charging bull into a standstill with the right brain electrode. It wasn't the first time *The New York Times* fronted a false report. Delgado had lied. Films of the experiment show that the bull did not suddenly become unaggressive and halt stock-still. Instead, triggering the electrode in its brain made it turn madly in circles, as if in a furious effort to knock off the electrode and the radio transmitter that controlled it. That's a reasonable response for anyone with a live wire in their head, and not at all a sign that special centers had been found and scientifically shocked out of action.

To see what's really going on in the feelings and changes associated with anger, we need to take a different approach, and for that we can do no better than to turn to the gentle English county of Shropshire, and look at one old man there, who seems far removed from the world of aggression, violence, and rage.

Anger Part One: Nerves, Brain and Adrenalin Rifle-Shots

Lord Emsworth slumped deeper in his drawing-room chair. He was alone, and his thoughts were wandering. His sister, Lady Constance, had finally moved and left him at peace; that Baxter, his insufferable secretary, had moved away too; and even prying Wodehouse, his chronicler of lo these many years, had finally stopped bothering him too, finally stopped beseeching details of this, that and the other inconsequential thing.

Now the lord could doze in tranquility, and deeper into his favorite chair he sank. His breathing grew louder, saliva dribbled from a corner of his mouth, and the muted gurglings from his belly were a sign that intestinal activity was underway. The pupils of the old lord's rheumy eyes grew smaller, and his heart slowed comfortably low.

But then, into this scene of dribbling bliss, there came a disquieting noise. A distant squeaking it sounded like, or possibly a scratching noise or a squeal. A SQUEAL! Could it be that Lord Emsworth's prized pig, the Empress of Blandings, was somehow in distress? Was being waylaid, manhandled, or generally messed about?

The lord jerked upright in his chair. Action was called for, vigilance, and if need be, attack. Suddenly, and in but a moment, the once de-

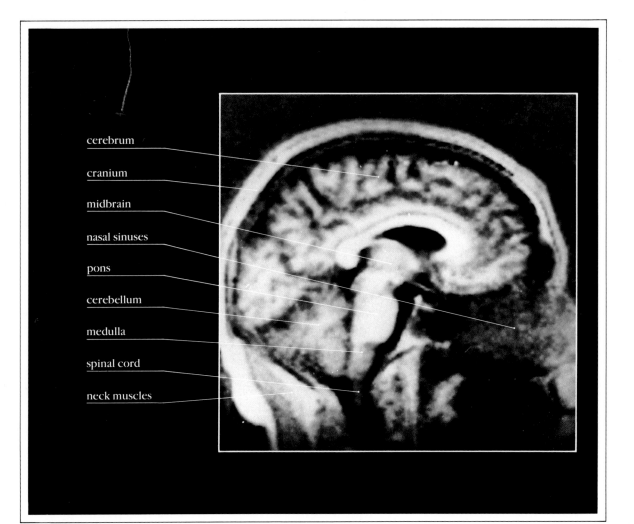

cerebrum

cranium

midbrain

nasal sinuses

pons

cerebellum

medulla

spinal cord

neck muscles

Brain levels as seen in living brain by high resolution NMR scan. The brain's gross anatomy suggests division into three levels. Right profile: advanced rational portion on top (cerebrum, filling whole top of cranium), emotional level in middle (midbrain, directly in center), and purely mechanical level on bottom (cerebellum for coordinating movement, bottom left; medulla for breathing control, lower middle). But this great oversimplification — brain levels intimately linked by constantly changing neuronal firing circuits — are too small and complex to be identified by most anatomists.

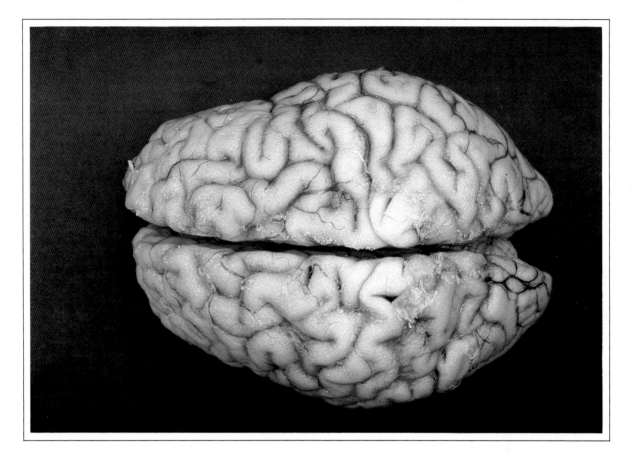

Left and right brains: some overall brain divisions do correlate with different functions.

crepit Emsworth was transformed. Saliva no longer dribbled from his mouth, and the pitiful guttural rumblings were no longer to be heard. Now Lord Emsworth's heart pounded fast with fury, his pupils dilated wide, and blood surged away from the useless gut, and into the muscles of his biceps and chest, now iron-hard and ready for all calls.

Fortitude, power, and élan were his. The spirit of the great lord's ancestors, who had held firm at Crécy, attacked in the Crusades, was triumphant. The culprits who had laid hands on the Empress would quail if they knew.

In a P. G. Wodehouse story the now fortified lord would probably leap upon the hooded hijackers of his beloved pig, only to find that they were the secretly returned Baxter and his sister Constance in disguise, who were making off with the sow for reasons of an at least thricely convoluted plot. But for our simple purposes it's enough to concentrate only on the change in the once drowsing lord. What wondrous transformation is this that came over him? What rejuvenation, switch, and reversal? Has it ever happened out of fiction — could it ever happen out of fiction?

The answer is yes. Inside each one of us are two incredibly convoluted networks of tiny cords, stretching every which way in your body. They're there to keep all your automatic systems running as needed — the digesting of food, the beating of the heart, regulation of temperature, and the like. Normally the two systems work in an even balance, and you have no reason to pay any attention to their useful but uninteresting churnings. But sometimes, and this is why we dredged up old Emsworth, they don't work in kilter at all.

All the symptoms an old man experiences slumping in his chair after dinner come about through one branch of this automatic system (called the parasympathetic by initiates) working to excess, with the other being largely switched off. It's your parasympathetic that has controls to slow the heart, increase saliva flow, and allow the elastic walls of the breathing passages to constrict, so making wheezing unavoidable. The opposite possible extreme state of the body, the fighting-ready, tensed and excited state, would come about when the second branch of your automatic control system (it's called the sympathetic branch) were switched on high, and totally overpowered the first.

What's interesting is that milder versions of this switch are capable of going on in us all the time. It only takes a slight overworking of your parasympathetic system to make you weak in the muscles, gurgling in the belly, and dribbly from the chin. This is in fact what happens every time you treat yourself and nap contentedly after a big meal. A mild blast of the sympathetic system however will turn you into something of a tensed, more powerful, minor Hercules. This is the state you're likely to get in whenever you find yourself really getting worked up about something, definitely starting to get angry.

Now this leads up to the fundamental question. What does it? What causes the switch, and makes one system cut down while the other is turned on high? If we knew we'd be a long way toward knowing whether anger and aggression are unstoppable sensations that just come over you, or whether they're under cool, rational control.

It turns out that the curious networks of the two automatic control systems have been traced back well into the brain. They're hooked up to specific clumps of cells in the hypothalamus, that organ the size of a finger joint located way back of your sinuses, near the bottom of the brain. (We'll look at it in more detail in a later chapter.) That link puts our question one level farther back. What is it that controls the hypothalamus: gentle reason? Or coarse emotions?

To some extent it's neither. A lot of what the hypothalamus does is fully automatic. It has, for example, more blood gushing through it than any other portion of the brain. Little sensors in the hypothalamus poke their wrinkled surfaces into that blood, like a bather warily dipping his foot in the pool, and if they register signs of too little activity, or not high enough a temperature, will automatically cause both automatic branches of the nervous system as well as various hormonal systems to get to work to set things right. That response is locked into you, and no one can do very much about it. Nor should anyone wish to: someone with a seriously damaged hypothalamus would find themselves choking, freezing, sneezing, panting, twitching, itching and in all ways coming to pieces even in a well-heated room on an ordinary day.

For the specific controls associated with anger though it's different.

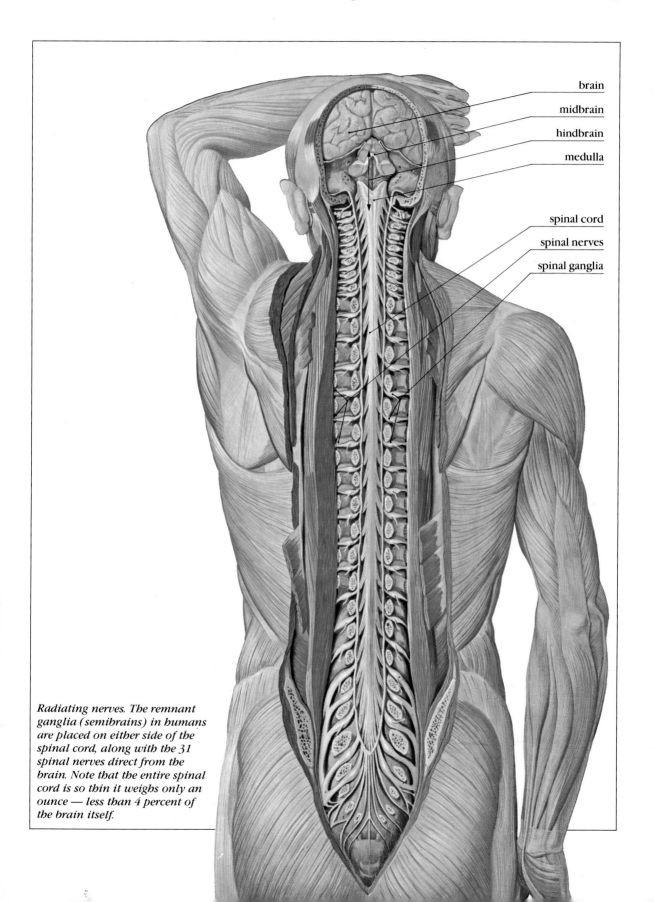

brain
midbrain
hindbrain
medulla

spinal cord
spinal nerves
spinal ganglia

Radiating nerves. The remnant ganglia (semibrains) in humans are placed on either side of the spinal cord, along with the 31 spinal nerves direct from the brain. Note that the entire spinal cord is so thin it weighs only an ounce — less than 4 percent of the brain itself.

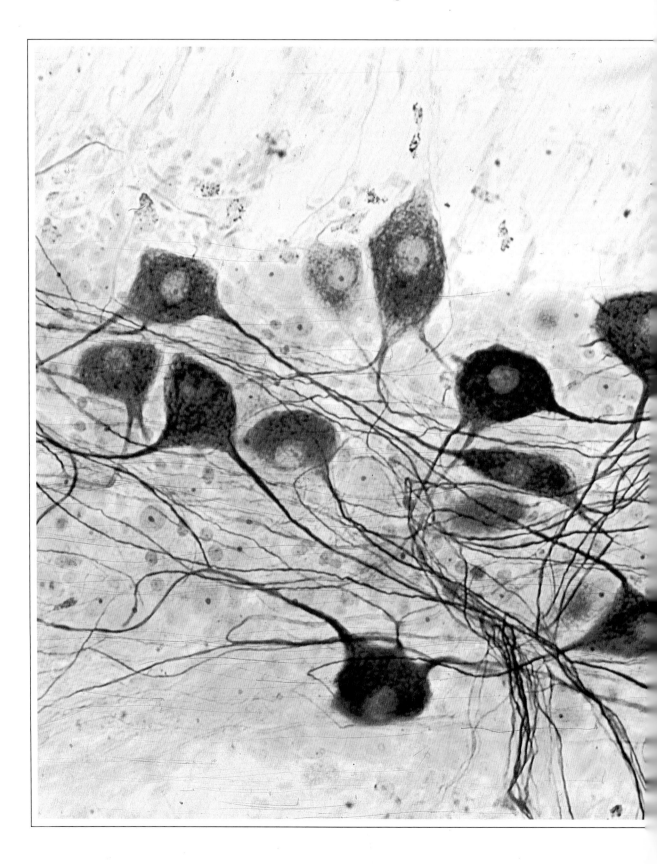

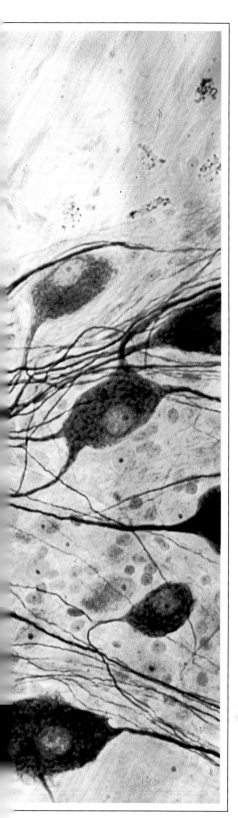

Quite a bit more is known about the volition behind a speed-up of the sympathetic system; a volition that's packaged there in your brain. The parts of the brain that can control the hypothalamus have themselves been found, and a look at one of the most important of these, the amygdala (Latin for "almond-shaped," which it is), brings us very close indeed to our answer.

Until quite recently that amygdala was thought of as nothing but a raw anger center. Trigger it in laboratory animals and they rage and hiss. Take it out of inner-city black leaders, various neuroscientists suggested, and tranquillity for white America would be ensured. Ronald Reagan even provided funding for the chief proponent of this view, back when he was governor of California in the late 1960s.

A second look at the tiny amygdala though has shown otherwise. Just because the amygdala fires when we're angry, that doesn't mean that it's the cause of that anger. In other parts of the body it's obvious — just because we have legs to run, no one would say that legs are what cause us to run — but here in the brain, with the leftover notion of Original Sin to cloud things up, that point is often forgotten.

The nerve fibers that stretch into the amygdala from deeper in the brain have themselves been carefully traced in the past two or three years. What's been found is astounding. Those controlling fibers come not from other emotional centers, not even from raw sensory input, but from . . . your rational thinking cortex and memory stores! That closes the circle. There at the start of it all is the cool, rational you.

This last link is the crucial one. Your brain is intimately constructed to keep your emotions under your own rational control. The links from cortex, down to amygdala, then to hypothalamus and the sympathetic system see to that. Just think about it: whenever you come across a situation that might lead to anger, ultraprecise electrical flashes are whirring around in the amygdala deep behind your nose to make the decision of how you're going to feel. Suppose you come out of the house and find that someone has broken off your car's antenna. Don't think dark thoughts about the delinquent who did it, and no orders will pour out to end up sending your heart beating faster, and your breathing deeper and mad. It's just as simple as that.

Sometimes though you do get mad. Sometimes it wasn't just the measly antenna that has been broken off, but the whole car that was trashed, that *is* being trashed, and you see your long-disliked neighbor's fourteen-year-old kid still at it. Rare are the dispassionate souls who will accept that with philosophical demeanor; most likely the feeling of anger will now come charging through you to the full.

How does this get carried out? What is it about this mysterious sympathetic branch of your nervous system, how is it built and where

Nerve cells from near the solar plexus, one autonomic ganglion among many — yet especially sensitive to a sudden impact due to its near-surface location and its radiating leads to nearby organs.

does it go, to carry out the anger activation orders when they do come down from your brain? The story that's been worked out is incredible. We've already gone up from body to brain; now let's go down again inside the sympathetic nerves themselves!

Once out of the hypothalamus, the commands of your sympathetic nerves pass into several dozen little white "semibrains" that you have scattered about on your back. Lower creatures have them too, from the 60-foot-long giant squid with its 10-inch headlight eyes, to the tiny bee, ant, and fly, but in those creatures they're much more under autonomous control, as the central brain is much less powerful to begin with. In humans the semibrains are knots of nerve cells that act as staging points for information flow from the brain and spinal cord on outward.

From the gnarled semibrains you have in your back, these living remnants of our lower past, nerve channels travel out to all the organs that the sympathetic system affects in moments of rising anger. These final nerves are firing to some extent all the time just to keep your body going; it's when they get the signal to fire in excess that the rippling emotional changes of anger are felt. Like most nerves they're really just squishy bags of protoplasm, with most curiously a long channel stretching most of their length. This channel looks like a hollowed-out bit of spaghetti, and filling up inside it is a two-way flow of protoplasmic goo.

From the main part of the nerve, special translucent baskets ready to be filled with transmitter chemicals are sent skittering along this channel down to the end. The baskets are delicate objects able to survive only two days before unraveling. To avoid an unwanted dismemberment halfway along the channel, they get flicked into special low-resistance speedways, or tugged along by tiny threads that get the baskets all the way to the far end in time.

Also floating down the main channel of the nerve, but in a much slower undulating wave that can take a week or more to reach the end, are all the items needed to see that the waiting baskets get filled. These helpers include lots of enzymes — batterylike power sources too — and when they contact some of the right amino acid dribbling into the nerve tip from the blood, set hastily to work to build up the substance noradrenaline (a chemical near-twin of the powerful adrenaline — we'll use the general term hereafter).

As soon as the adrenaline is made a few more enzymes cart it over into the open waiting baskets. It's important that be done quickly, because the raw adrenaline is so potent that it would start bubbling away and dissolve in the cell if left loose. Finally, once enough adrenaline molecules have been packed in, the basket lifts up its own cover, and starts to snap it closed, like a hungry but slow-motion clam clamping its shell down over a meal. The hinge-closing basket is not fast enough to keep a few adrenaline molecules from slipping out, and a few enzymes, electrified calcium, and other odd bits from the main nerve go slipping in. But it does end up with 10,000 to 100,000 adrenaline molecules firmly locked inside and for what's going to come any moment now, that will do just fine.

This carefully prepared-for visitor is an electrical impulse, that started up in the brain, flashed through the switching point of the semibrains, and is now hurtling through the nerve. This impulse is so fast that even the high-speed channels bringing the baskets down to the nerve end seem sluggish in comparison. Those channels were lucky to reach 200 mm a day; the electrical charge zipping along the nerve now can surpass 100 mph on the broadest straightways.

Thundering into the nerve tip the electric signal packs an electrical charge of 70 millivolts. That isn't much compared with the 220,000 millivolts of ordinary house current needed for an electric shaver or food blender, but here it causes the carefully packed adrenaline baskets to carom about so drastically that many of them pop clear out of the nerve's tip. (A few miss their chance and go to a waiting pool, where if still not used in a day or two the nerve turns on them, attacks with enzymes and dissolves them.) The released baskets split open as they come out of the nerve, and send all the adrenaline molecules that were inside them pouring across the small gap separating the nerve from the organ that it's targeted to affect.

The loose adrenaline molecules have so much momentum from the electrical signal that fired them, that they tumble across that gap in a fraction of a second. On the other side to meet them is an incredible object. The cells of the target organ all have a semifluid membrane wrapped around them, and in them now, just across from the nerve, there are humongous iceberglike proteins, bobbing up and down in that membrane.

Luckily the bobbing receptor is forbidding only in distant countenance. Seen up close it has a tiny section on its top where the transmitting adrenaline molecule can comfortably fit in. When enough of those niches are filled, the bobbing iceberg deftly opens big gateways that it's connected to in the membrane of the target cell, or sometimes even shakes loose a second messenger lower down in the cell. With that last step, the anger thoughts that started up in the brain are complete.

 The results are tuned just right for each different cell it is that's been hit. Adventurous leads of the sympathetic system that stretch into blood vessels in your arms are quick to make them tensed and strong, those wending into the muscles controlling your lungs indirectly cause them to speed up your breathing, and fast. Other sympathetic leads stretch into your heart, there to speed it, into the skin, to heat it, and indeed to all the other parts of your body that undergo changes when you start to get worked up. There's more. Much smooth muscle gets contracted wherever the sympathetic nerves hit, while the anus and other sphincters, however, suddenly squeeze closed now; digestion, including all 35 feet of your coiled small intestine, switches off as well. There are more important things to concentrate on now!

All this will be happening if you start fuming in a traffic jam when you're late for work, or if you come out with a long-simmering argument with an office colleague once you're there. The incredible release of the little adrenaline baskets are like precise rifle shots — millions of triggering pinpricks — all over your body, going to just the

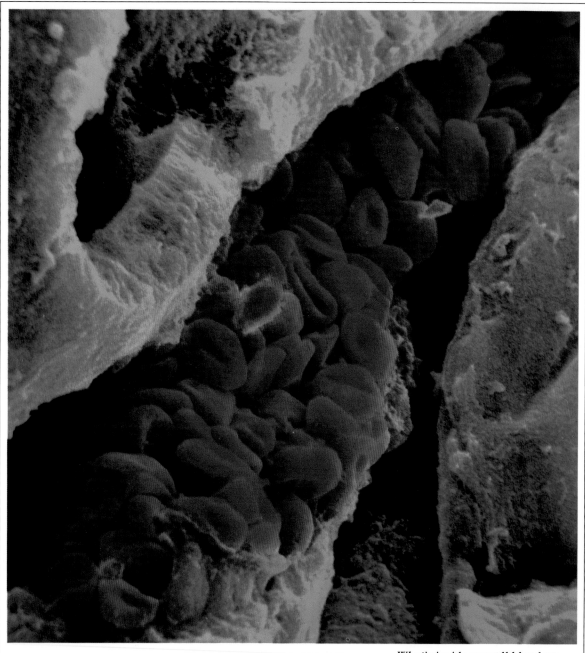

What's inside a small blood vessel: vast phalanxes of red blood cells on the move. Scattered throughout them in level 2 anger will be adrenaline molecules from the suprarenal medullas; a more energy-efficient way for the body to distribute adrenaline than through the scattered sympathetic ganglia.

right places to make you the most possible opposite from that rheumy old man we started with, slumped in his chair, passive and weak. Now all fired up, you're powerful, ready and strong — a consequence, we have seen, not of blind emotion, but of the rational parts of your brain switching your sympathetic system into high.

Anger Part Two: Hormones, Body and Adrenaline Shotgun-scattering

That switch up on top comes out in another way too. Think back on the tiny white semibrains, that worked as staging points in the firing of your sympathetic system, and which you have scattered quite neatly all along your back. You could imagine one of those dispersed nerve knots growing and growing until it was hundreds, thousands of times bigger than the others, with labyrinths and crannies able to store up vast hordes of adrenaline baskets to send out when the signal comes. The organs and muscles of anyone so afflicted that respond to adrenaline would really get a surging kick when he got angry.

The growth that could do this would be an awful sight, a deformation of one of the normally tiny semibrains that would have had to begin back in the womb, then wildly gorge and grow until it reached its final, horribly swollen size. Developing from a semibrain, it would be bulging somewhere in the back or stomach where the other semibrains are. You might feel sorry for someone with such an unfortunate growth, but that would probably not be as strong as the shuddering repulsion you would feel, and the thanksgiving you would silently give for not being so afflicted yourself.

Well, if you're of the shuddering sort you might wish to sit down, for the person afflicted with that swollen anger semibrain growth is . . . yourself. You've got it, two of them in fact. Both are perched over the kidneys in your lower back, both grew out of your normal semibrain nerve knots when you were in the womb, and both are hundreds of times bigger than the usual ganglions that trigger an adrenaline release. They are called your suprarenal medullas, and when they get the message to let loose their adrenaline stores, watch out.

Within each one are millions of minute baskets filled chock-full of adrenaline or noradrenaline. Just as in the sympathetic nerve endings, these potent molecules are only held in by the Saran-Wrap – like outer covering of each individual basket. It's a precarious setting. Splashing right through this adrenaline warehouse are sheets of cascading blood pathways just waiting to rush them away. As long as the baskets stay tightly sealed inside their cells, you're OK. The blood can gush by without picking up a drop of the purified adrenaline. It's when the Saran-Wrap–like basket coverings split open that everything changes.

Almost all the adrenaline tumbles straight into those rushing blood waterfalls. And what sets the adrenaline free is — you guessed it — just one particular set of leads among many from the stretching nerve fibers that make up your so carefully controlled sympathetic arousal system. When a lot of adrenaline is needed, it's cheaper for the body to get it this way than to laboriously charge up each individual nerve to set its own relatively paltry amount loose. The latter is like expensively phoning every member of the company when there's some fresh news to spread; the former, like simply running off a memo on a central Xerox machine to distribute.

The freed adrenaline from this superstore in your back heads out with the cascading blood, curves through a prearranged tunnel heading up to your heart, and goes in there to get pumped out into the whole body's circulation with each squeezing heartbeat. There it enhances all effects that the sympathetic nerves have started, keeping up all those changes in the body that we've come to feel are an integral part of feeling mad. Let's take a look at a few of them in turn.

HEART   This is the organ most people continually mislocate, being actually based largely on the center-line of the body (a good third of it is on your right), with only a small part stretching over the 3½" to the left side where it pokes up toward the surface and is so easily, and misleadingly felt. The half- to three-quarter-round hollow muscle of the heart is also the organ we feel most strongly just at those moments when we think we're most fully living: when excited or in anger, when thrilled or having sex. Especially among men, the heartbeat has become a tribal fetish, endowed with magical powers that ensure life and strength, an object vying only with the penis as a source of near-hysteric concern.

Individual adrenaline drops from the sympathetic nerves triggered the precise bit of muscle or control channels they hit in the heart to go more strongly. The loose adrenaline coming out from the medullas doesn't go to the exact same spots (the sympathetic nerves have vacuum-cleaner-like devices to pull in stray adrenaline from their target) What it does instead is tumble over and over along the rest of the heart muscle, until, like a rolling sagebrush hitting a cactus, it thwacks into a receptor projecting out from another unoccupied part of the heart. Binding electron clouds form to make sure the new arrival sticks tight, and then the adrenaline has no choice but to get to work.

Heart cells are sensitively constructed to beat with greater vigor when such loose adrenaline arrives, and as the medulla has released a bushel that's enough to keep up the pounding that the sympathetic nerves started. Blood gushes out of the heart at higher speed now, distending the elastic walls of the aorta even more than usual when it hits, and setting up incredible turbulence patterns. There can be eddies in the blood, whirls in the blood, as well as vortex shedding, fundamental tones, and even high-frequency harmonics within the breast of the now storming angry person. High friction builds up between the successive layers of blood being so turbulently pumped. That raises the apparent stickiness of the blood in the arteries, which makes pumping it forth even harder, and helps raise blood pressure throughout the entire body.

Little of this lasts for long, as half your loose adrenaline is dissolved apart by the blood in its first sixty seconds on the town, but it will be a large part of the heart surge you get when finally you do decide to let loose and tell that insufferable office-mate of yours just what it is you think of his finger-cracking. *And* his hogging of the pencils.

LUNGS The change here is, dare one say it, breathtaking. Sitting calmly in an airplane seat as you wait for your shuttle flight to taxi for takeoff, you are likely to be breathing in perhaps 1½ liters of air a minute, about three small milk cartons' worth. Now let a precocious juvenile sitting in front of you suddenly undo the latch on the emergency hatch in a fit of scientific curiosity. The adrenaline surge that you get as you think foul thoughts about jettisoning the miscreant through the opening, which now has to have its door time-consumingly refitted, is likely to send your lungs flapping madly enough to bring in perhaps ten times as much air a minute, enough to fill a medium-sized filing cabinet.

A simple speeding up of the lungs couldn't bring in those gulping extra liters alone. Thirty percent of the air you breathe goes right back out without having been touched, because it never gets pulled all the way to the working surface of your lungs. The breaths now have to go deeper to get their much larger uptake. The free-ranging and mischievous adrenaline, even acting on muscles below your lungs and around it, is only too willing to oblige. The now suddenly faster lungs also get a helpful fine-tuning. There are always ready brain sensors to speed things up when carbon dioxide is building up and the lungs' pumped-out blood is too acidic; there are also sensors speckled onto the inside of the aorta that have indirect controls to speed breathing even more if oxygen levels in the blood have fallen too low instead. Special blood pressure receptors you have reassuringly stuck in your neck at just about chin level also do their bit to make the adrenaline surge work best.

LIVER Not far below your right nipple is the three-pound blob of your liver. Stored blood sugar is locked away in there, stitched into glycogen molecules and wedged deep inside various cells. When you get mad loose adrenaline from the medulla storehouse in your back rolls and bounces over the surface of those cells, and, just as in the heart, smacks into prearranged receptors that here start working to bring that useful sugar out.

One chemical messenger floats down into the cell from the adrenaline-filled receptor, another takes over and leads to unfastening the sugar that's been so carefully joined, and soon the molecules of sugar, thousands of billions and more of them, come tumbling out of the liver that the triggering adrenaline so recently entered. The sugar mixes into the bloodstream, and streams with it to the brain, there to give it the fuel that can make it tensed-up and more efficient. (A good deal of stored blood sugar is also lying about in the muscles directly — the glycogen we touched on in chapter one — and the adrenaline gets that ready for use too.) Students before an exam can get this effect though it will be of no great use; professional football linesmen during a game will get it and use their muscles at maximum output so much accordingly that they can end up with much lowered no blood sugar, stored or otherwise, in their body when they're done. If that's not an excuse for collapsing flat on a training room table afterward and refusing to pipe up to Howard Cosell, nothing is!

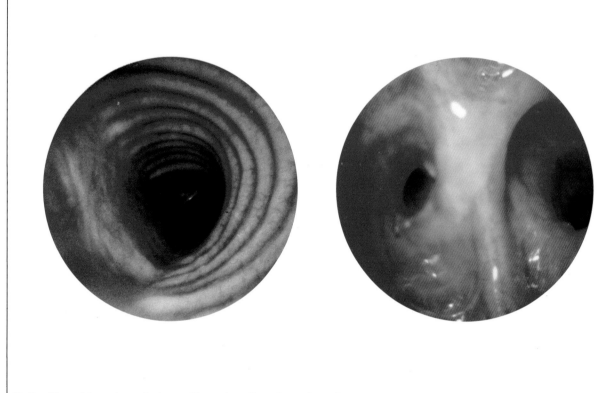

*Path of inrushing air — during ordinary breath and panting of anger
alike. Left, down trachea (breathing tube); center, bifurcating to either lung;
right, arriving in densely packed minute chambers of the alveoli where gas
exchange takes place.*

FAT All the fat in your body can be considered as a single spread-out
organ. The main parts just happen to be in the hips, belly, or wherever. Fat cells are bulging spherical containers, loaded with a single
big drop of pure, unadulterated, shiny white fat. They're also active
metabolism sites, with an estimated 10 percent of their contents being
shuttled out or shuttled in afresh each day; a transport helped by the
terrific number of blood vessels you have wrapping around every
bundle of fat there is.

Fat is so carefully arranged in you because, on top of being a good
insulation, it's an incredibly efficient storage form of easily available
energy. A lot of excess blood sugar is automatically converted by your
body into fat, and on the chemical level that's not a difficult project at
all. Sugars are carbohydrates: the most abundant group of biological
compounds on earth.

Loose adrenaline from the medulla that wasn't used up in your
heart, that didn't get caught in your liver or around the lungs, is almost
certain to get swept by the bloodstream head-on into a healthy glob of
fat. If the blood takes it to the abdomen, fat is there waiting, and ditto
for even the knees (what's called popliteal fat), the groin (inguinal fat)

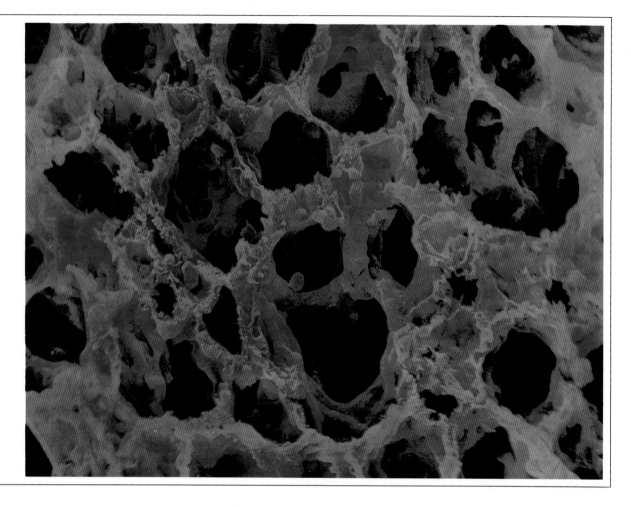

almost everywhere — there's no escape from the bulging spherical fat.

When the adrenaline hits, it starts the carefully built up fat back on the undressing path toward the simple sugar from which it began. Quickly on the edge of the fat cell the contortions happen: carbon pairs get ripped off, something like ordinary soap gets prepared as one intermediate step, and finally out bubbles a stripped-down fatty acid, all set to be carried by the blood to the muscles and help in the sugar-needing energy processes there. When someone's fatty jowls tremble in rage, it's not just on the outside that the shakeup is going on.

All these bodily changes produced by your internal adrenaline surge at work are crucial in giving you the feeling of being mad. In fact you could even say that without them you wouldn't be really angry at all. Suppose that your home computer suddenly printed out the statement that it was fed up with being treated as a dumb machine, and that unless you treated it with more respect it was going to get angry and fume and refuse to compute. You probably wouldn't believe it. Even if the computer repeated its assertion, typing out "I'm angry, I'm angry" this time with a string of emphatic exclamation marks trail-

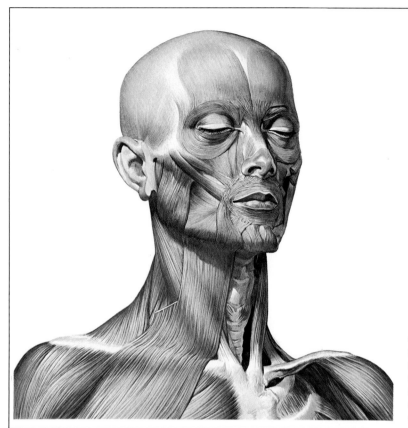

Under your skin: the supporting muscles of the human face. Responding the same way in everyone, they're the major reason for the constancy in facial expression of emotion in widely different cultures.

ing along afterward, you probably wouldn't give in. The thought that someone else in the house had been tinkering with the machine would be far more plausible.

Now imagine you saw someone standing by you with his arms taut and fists clenched, his face red, veins throbbing in his neck and forehead, and panting with his lips curled back in a vicious snarl. Here by contrast if he said that he *wasn't* angry, panting it out through clenched teeth, you almost certainly wouldn't believe him, and might even back away. These body changes are a good part of what angry emotions are all about.

Beyond Ordinary Anger

Beyond the two-part action of ordinary anger are a number of tenuously charted further realms. The bizarre link between anger and sex is an all too common one. A lot of people get sexually aroused after a spousely spat, and end them assiduously in bed; there seem to be a good number of people who get aroused by direct violence too. The violence/sex connection of popular mythology is especially clear in slang.

Possibly it's because both sex and anger can be feelings that seem to come over a person, making things happen automatically and without the usual burden of full consciousness. There's some support for this linguistically, as consideration of the vicissitudes suffered by the Latin word *rapere* can show. Literally the word means "to be seized," or "to be under the control of." Going one way it has turned into the English word *rapture,* but along other lines it served also as the base for the word *rape.* The feeling common to both is of being transported by an extreme emotion, be it anger or sex, and feeling that you're held under its will. Viewed that unfortunate way, the one will serve for the other.

Lawrence of Arabia typified the connection. Personal letters long suppressed by his estate show that the mighty Lawrence was fond of not just removing his bedouin cloak when sunset came over the hills of his Middle Eastern haunts, but preferred to go on to slowly remove all the rest of his clothes and then lean forward, while a similarly inclined comrade in the fight for the Arab cause beat him angrily with a stiff whip until he ejaculated.

In that extreme form it's rare, but to a lesser extent it's a linking, an impulsion, that's only too widespread. In a study of over 3,000 college students in California and Canada carried out recently, 35 percent of the men said that they would gladly rape a woman they met if they could be sure of not being detected. Thirty-five percent!

Another unfortunate direction normal anger can develop is that of blind rage, anger uncontrolled. Here the gradual stepping up of the body that we've seen in the sympathetic and medullar responses start losing all their worth. Muscles can lock, while blood vessels near the postage-stamp-thick retina on the back of the eye can swell and buckle and shift to make the sufferer actually see red. None of these changes are of any particular use. Indeed, above a certain speed your heart pumping faster has no chance of producing ever-increasing strength. Its internal valves, normally delicate sails of fibroelastic webbing, just get stretched and buffeted out of shape by the rushing blood torrents, and the thick outer heart walls strain to keep up the pumping, like a car's piston flapping uselessly in the wrong bore. Less useful blood is sent on than a lower setting might produce. The speedup just intoxicates the sufferer to continue in his furied state even more.

Perhaps above all, the furthest realm of anger is beyond simple physiology entirely. The ancient Greek philosopher Aristotle, writing 2,400 years ago, recognized this. "The angry man feels pain," he said, "but the hater does not . . . . Much may happen to make the angry man pity those who offend him, but the hater under no circumstances wishes to pity a man whom he has once hated; the angry man would have the offenders suffer for what they have done; the hater would have them cease to exist."

The same mentality, the same clear, level-headed evil choice, is behind many of the civilian slaughters our twentieth century has been so profligate with. When soldiers of the German Das Reich division fought members of the French resistance in mid-1944 who were harassing their advance to the D-Day beaches, adrenaline surges and direct visceral anger were probably going strong on both sides. But when the same German soldiers stopped off in the village of Oradoursur-Glane in France one sunny June morning, and herded hundreds of

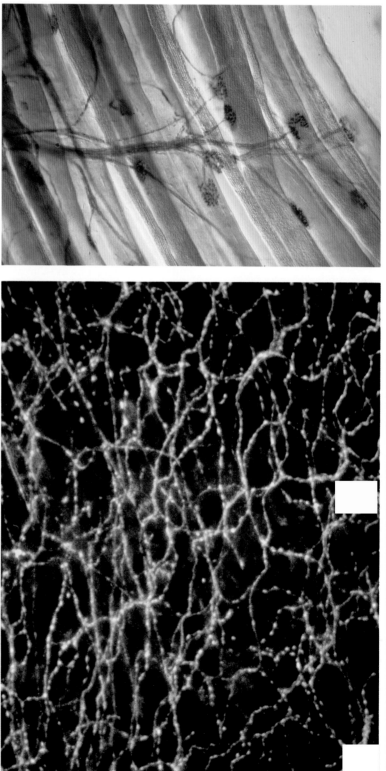

The adrenaline continuum. Muscular tensing in even happily excited gesturing is associated with nerve channels that snake throughout body to release muscle contracting neurotransmitters at end branchings (above, right); in case of nerve pathways from sympathetic ganglia (semibrains) scattered down back this release seen as net of green-glowing (nor) adrenaline droplets. Entire network of endings (lower right) might fit into and control just one .004 inch segment of artery feeding muscle — thousands more release-networks this complex are needed to support a single extended finger.

town children into a church and then set off an enormous, murderous fire-bomb among them, that was a cool decision beyond what our scientific physiology, so far, can possibly hope to explain.

Let us leave anger hanging unknowingly there, and turn now to the other side of the great basic emotion divide: fear. We will start with one of the most unpleasant examples of acute fear, and gradually work our way along to fears both historical and diffuse.

Raw Fear Example: A Mugging

Muggings are painful and humiliating and they happen fast and they leave you a wreck. They also happen more in big American cities than anywhere else on the globe, and according to FBI figures have a good chance of including *you* in the statistics sometime before our soon-to-finish century is out. And when that mugger gets you, the things your body will do are startling to behold. Beginning, fairly enough, with a startle.

Not an ordinary startle. Not a nonchalant flick of the head sideways, eyebrows raised in polite attention. No, let a murderous shout of "Hey, freeze!" ring out unexpectedly from behind you in a dimly lit basement of the company parking garage, and then, if the scene could be recorded with an ultra-high-speed camera and played back, the following uncanny actions would be seen.

The shout emerges from the mugger's mouth at 600 miles per hour. It's a silent scream at first, for only after .01 second will the first air vibrations of the shout reach your ears. Those vibrations set the outer flap of each ear shaking; that flap acts like an electronic aerial to capture the sound and redirect it around a sharp bend then straight on down into your ear canal (earhole). Still moving at 600 mph, the shout travels in a plane wave down the approximately one-inch funnel of your ear canal, passes over the sweat glands (closely related to armpit sweat glands) in there, skirts among the boulders of wax and finally cracks straight onto the bouncy membrane of the eardrum and thence in toward the brain.

A single .04 second later the startle reflex proper begins. Your eyelids slowly, flutteringly, begin to close, the lashes bending back in the wind from the move. Then the head starts laboriously tipping forward, and the outer tips of the shoulders tighten in rippling muscular waves and begin to mysteriously rise up in the air. The abdomen flattens inward, as if in a deep, powerful breath, while the entire torso buckles forward in an impossible, ridiculous, formal tuxedo bow.

Each one of your fingers starts to unerringly straighten out from the hand, and the elbows start accelerating in a drunken, unwieldy flight away out from the body. The entire spectacle wheezes slowly lower to the ground, as the kneecap pushes forward, the whole joint of the knee goes with it, and the legs get cranked angularly in the direction of a pre-hopping crouch.

Everybody does it in exactly the same way. Even a seven-month embryo in the womb will jolt out the main points of the startle response if it's stroked with a thread over the eyes; even a fervent terrorist holding an AK-47 can let it drop as the startle reflex starts him flailing, as

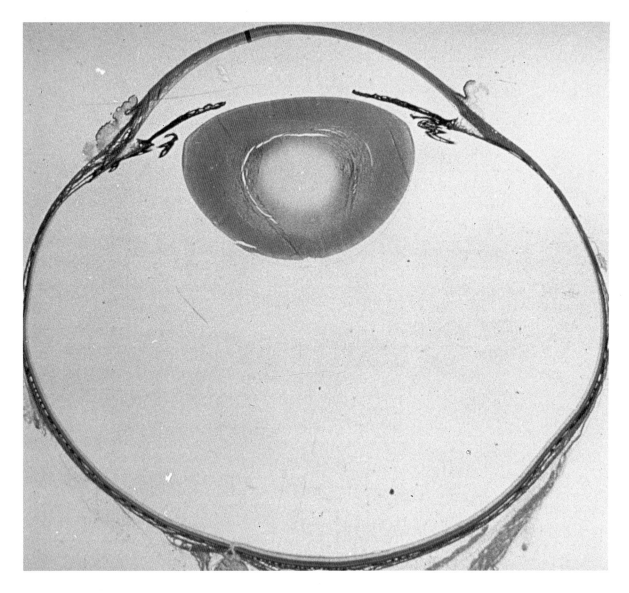

indeed the British SAS demonstrated when they supplied incredibly loud flash grenades to help storm the Lufthansa jet being held at Mogadishu in 1977. This is amazing. Stretched all throughout your body are a tangle of nerves, some leading to the muscles that poke the fingers out, some dangling down to the muscles that pull the knees into a crouch, which wait silently going about their normal business, sometimes for years on end, until a sudden noise or fear calls out a firing mechanism in the brain to coordinate them in this unbelievable dance.

Mammalian eye in cross section: lens (at top) will slightly buckle when ciliary ligaments holding it decontract under extreme emotion; semicircle region in front of it can be compressed enough then for small amounts of sticky fluid it contains to be expelled out of eye through edge of cornea.

After the startle, the color. Even while the last twitches of the startle reflex are still going through you, even before you've managed to turn around and see the mugger, your harmless cartilaginous nose starts to fade a mysterious cool-glowing white. The vagina joins the shift to white, the stomach joins the shift to white, and as for the face, well: not only does it transform to a spectral white too, but it also goes clammy and cold — the famous cold sweat — due to a rush of perspiration that has mistakenly broken out there even though the color change has ensured that there is little hot-blood-filled skin left exposed to cool.

A cloud of fresh smell molecules starts drifting away around you, as many billion of newly minted chemicals fall through the blood vessels into your sweat glands and evaporate in churning gusts off free into the surrounding air. The heart speeds up, contorting and twisting in the middle of a beat as its special accelerating centers sound the call, and the saliva tunnel under your back molars slowly squeezes and halts the flow that had been coming out.

Some of these first-moment changes are good for you. The weird startle bending is a useful way of protecting your inner organs, and the compressing belly it produces even squeezes the abdominal organs enough to squirt useful high-pressure blood up along the veins into your chest. The ghostly color shift too has its merits, leaving more blood for the inner muscles, and lowering the bleeding from any superficial cuts that might come.

Other parts of the first mugging changes — the radiating fear molecules, the mistimed cold sweat and vagina paling — are not of any noticeable use. But all show the body doing what it can, using sympathetic nerves to the heart here, command structures in the brain for the startle ballet there, to help prepare you for the next moment, the ordeal itself, when you do finish turning around, and face the man who is about to mug you straight on.

Slow down the camera even more and an amazing spectacle now comes to view. Facing the mugger there in the garage, your elbows float back out, your shoulders start to lower, and all quivering begins to end. The ponderous, fluttery eyelids shift back up to where they had come from, tugging the lashes the other way now in the wind. That last move reveals the vast orbs of the eye, and they, after some preliminary straightening and focusing, start on an unseen odyssey across the mugger's body. The great skating eye escapade has begun.

First the whole eye starts spinning slowly in its lubricated socket. Gradually it picks up speed, tugged by now-tautened muscles that thread through useful capstans hewn in the very skull beside it. Top speed is soon reached, and the now-whizzing orb is on target straight to the most captivating of goals: the center of the mugger's eyes themselves.

There the spinning eye stops, holds frozen in a transfixed look for a .28 second that's near universal in fear, and then starts off on its rolling free glide for more. All this time the rest of the body is still, frozen motionless in brief terrified anticipation of what is to come, and has not been even properly viewed or registered yet.

Across the terrain of the mugger's body the frantic eyeball spins, retina switched off, registering nothing, until with a powerful deceleration it jerks to a squishing halt and switches on its full searchlight gaze once more. This second way-station peer is near certain to be of some outline of the mugger's body, the corner of a jacket perhaps, or

maybe the contour of the shoulders. A stationary glance, a peek, and then another rolling turn is on, switching the eyes to some other corner, some other visual drink.

The grace of these glides is enhanced by the duet they are joined in, for the spheroids of the mugger's eyes are whirling and decelerating too. It's two briefly frozen bodies, four frantic moving eyes, engaged in their instant reflex visual assessment. All are in line for an uncomprehending .28 second at the first stop, then veer and weave to the other's passive outlines thereafter.

Only slowly does the panicked brain register what is happening. Even before the body unfreezes, even while your eyes are on these incredible glides, a picture is entering your brain, but it's an unrefined picture the brain can do nothing with. In the thick nerve bundle heading back from your eyes (the optic nerve) the first image is a simple assemblage of dots. A TV monitor hooked into that nerve would show something of a very coarse image of the mugger.

A few inches deeper into the brain the optic nerve branches. A TV monitor getting its pickups there would show two separated images of the mugger, split vertically down the middle now as if seen through a split-image camera rangefinder that was totally out of skew. This grainy double image of your attacker reaches the middle of the brain before the second eye glide has even begun, but it's still not ready for assessment, let alone near the conscious centers that can clue you in to what exactly is happening out there with your foe.

Further into the brain this incoming visual image falls, weaving its unerring way amidst the tangle of ten billion brain cells, none more than a hundredth the thickness of a hair, and all crackling live with flashing electrical charges. The proto-vision sputters at full speed all the way to the very back of the brain, and only skids to a stop a bare millimeter from the edge, just under your skull a bit below the point where men first start going bald. There the image splays over a region 5,000 times larger than what it was at the tiny retinas in the eyes. It also gets split up and processed in a vast number of tiny nodules that float on the surface of the brain. Each nodule is made of six or more separate levels, like floors of a house, and all have intermediate trapdoors to shunt the signal up and down. The trapdoors aren't useful all the time, but seem to be held open and then slammed for processing in discrete intervals of perhaps $\frac{1}{5}$ second each.

A TV monitor connected here would finally lose the mugger's image in a hopeless blur, split into thousands of little sections to match the signal in each nodule. These TV fragments would not even be selected close-ups either, but just abstract codings of the angle, shape, or size of the image in that particular part of the visual field. In the brain however this coded blur of vectors and coordinates is exactly what the consciousness centers need to register and finally click on. This is the only way each cell can get a manageable portion of the image to work with, and when that happens, you do finally see who it is that's facing you. Now the real awareness of fear begins, now not only is your heart pounding but you see who it is that's made it so; who called and terrified you and now perhaps is going to come in to make it worse.

That's not all. At the same time as this jolt of consciousness takes place a second copy of the mugger's image is being sent not up to consciousness, not even down to the reflexes, but instead skirts back to our old friend the amygdala, and from there enters a part of the brain that puts the scarifying image promptly into stable memory. There the image will remain, nestled amidst the other notable memories of graduation days or first loves, immune to anesthetics, cold, sleep, and the passage of possible decades to come.

Those leads from amygdala to memory input during fear are so deeply ingrained, that if they happen to be snipped your brain will send out new channels and develop fresh cross-paths to see that the same branching into memory stays ready. Whatever happens next between you and that mugger, your body has seen to it that you'll remember for a long time how it all began.

Lie-Detecting Interlude

Is that it? Fear on the one side of a rousing sympathetic system, anger on the other, with not much of anything in between? Well of course not. Most of the time we don't get either mad or angry when worked up, but instead bluster, bluff, chortle, stammer, or do any one of a host of other common compromises that we muddling humans fall into more often than not. The whole notion that it's got to be one or the other, that under tension your body prepares you exclusively for fight or flight, stem from some theoretical musings of one Harvard Medical School professor, Walter Bradford Cannon, writing shortly after World War One.

At that time it was still a big political question as to whether or not the United States should have sent troops to France. Much of the Harvard faculty, where Cannon was, had been for war. Either you stood up and were tough, as Harvard's bravest so unquestionably were in the annual football games before the raccoon-skin coated crowds, or else you were a coward who slunk away and got clobbered, perhaps hopefully as the Yalie squad would do in the next seasonal finale. In that distinguished setting it was only natural for Cannon's carefully qualified ideas to be taken up in a crude and simplified form; for the idea to spread that whenever adrenaline started getting pumped through the body, it was to prepare the system for either anger or fear, fighting or fleeing. It was very much a macho science, from that impeccably macho setting.

Matters may have rested on that theoretical plane, but for the fact that the large businessmen who dominated the United States then were grabbing onto every possible pseudoscience available to justify the repression of workers, and the restriction of immigrants from overseas. (Some restrictive skilled labor unions were for this too.) The dubious field of eugenics, for example, was used as partial backing for the developing of IQ tests, which in turn were used in the early 1920s as a support for keeping cheap Italian, Polish and Russian labor from reaching the United States. The same intelligence tests were also applied to keep workers' education inferior, and to limit access to the elitist, reactionary universities.

Girl fascinated with snake — clearly an artificial pose; a nineteenth-century French print representing a sorceress, emphasizing how unnatural liking of snakes has been assumed to be.

In that distinguished setting it was only natural to misuse Cannon's-idea for the development of the social control tool of lie detectors.

Low-cost blood pressure monitors had been on the market for years, and it was the easiest thing for a tinkerer to rig them up with an easy-to-read dial, and proclaim that by measuring the working of the sympathetic nervous system they had invented a foodproof detector of lies, that would help safeguard established America from decay. Police stations used it (the Berkeley, California, police seem to have the honor of being the first), then some office managers expressed interest in it, and soon the chance for official respectability came when someone at the Surgeon General's office in Washington, D.C., was thought to be stealing small items, and one of the newfangled lie-detector specialists was called in to find out who it was.

There were many employees at the Surgeon General's office, but as only the most menial workers were thought to be potentially culpable, seventy mild-mannered black messengers were soon marched in to suffer the test. As the administrator of the lie detector reported of his examination later, "All subjects were negroes of emotional temperament, and blood pressure variations in all tests were great."

This is not surprising. Washington in the 1920s was still a legally segregated city, with U.S. congressmen and Supreme Court justices, among others, thinking it preposterous that blacks might ever be let into the restaurants where they ate. American whites hadn't let American blacks fight with them in the recently ended first World War, and were indignant that the French had, as well as found cause to award them 171 medals and the Croix de Guerre. The sheer shock for the messengers of getting a blood-pressure device wrapped around their arm and pumped up must have played at least as great a role as potential guilt in producing the cool notation that "blood pressure variations in all tests were great." Still a culprit was chosen — even though controlled English tests today show that accuracy in such a setting was quite likely to have been worse than random chance — and to lie-detector specialists today, that operator at the Surgeon General's office is considered an honored founding father.

As better technology after this early start made it possible for "lie detectors" to monitor more readings, things had to be found for the machine to scientifically measure. One researcher suggested measuring whether or not the subject looked away when asked an important question, while others said no, you had to check whether his breathing was faster. A dip into the repertoire of sympathetic nerve responses led another pioneer to remark that wouldn't it be a good idea to use the machine to measure whether the culprit's hands got sweaty, while another said OK, but how about checking whether he paused longer than usual in answering questions about potential misdeeds.

A parody of science on the march it might seem — but those suggestions were precisely the ones that have been incorporated in the lie detectors used by the 3,000 operators, trained at over twelve colleges, who give the several million annual lie-detection tests in America today. True to their origin the tests are still used disproportionately on unskilled women or blacks in menial jobs, and still suffer from the

The startle reflex is subdued in a trained soccer team about to receive a penalty shot, left.

Startle reflex in amateur soccer club (note ball hasn't touched them yet — ones doubled over closest to its approach when picture taken, below, show most classic response, including doubling over).

key error of not realizing that a given speed-up of the sympathetic nerves or other automatic responses can mean any one of a whole host of things — anger at having one's honesty questioned being a prime one. Valid applications of physiology exist; the business of lie detection is not one of them.

Dispersed Fear: Depression, Brain Chemistry and Diagnosis

What about when fear crops up in more ordinary life? How do the physiological correlates compare with those of acute fear? Everyone knows how dispersed fear can sometimes turn everything around you, all the mundane objects of daily life, car and office, friends and family, into looming, exhausting, unwanted presences. There is a sadness, a gloom, an intangible uncomfortableness in everything. It's what Churchill called his "black shadow," Pascal his "emptiness," and Louis Armstrong, so feelingly even in the midst of his apparent success, his "blues." Now it's called depression.

Temptingly, there do seem to be some simple brain mechanisms to explain it. A lot of recent researchers think that depressed people get that way when the chemicals that transmit nerve impulses in their brain go out of kilter. Since one nerve cell in the brain can have branches snaking out of it that end up controlling 250,000 cells in the body, that shows what can happen when there's even a slight chemistry problem up there.

Valium, for example, reduces the usual gush of inflowing chloride pieces that help carry nerve signals by making the brain's small amount of natural plugs against it (a complex substance called gamma-aminobutyric acid), work more effectively. Thorazine, for more severe cases, clogs up the triggering niches of brain cells involved in other kinds of information flows that otherwise excess amounts of the signal transmitter dopamine would be able to mess around with. It seems so reasonable, to so many doctors and patients, that an estimated 40 million Americans use these mood-changing drugs each year, with 240 million prescriptions being written, and Valium alone accounting for nearly half a billion dollars in sales.

But there's a catch. The more these drugs have been used, the more reports have come in that they really aren't working at all as they're promised to. Depression isn't cured, and a number of rather nasty side effects crop up in their stead. Even the apparently mild Valium, according to figures from the U.S. Drug Enforcement Administration, is behind more hospital emergencies and drug-related deaths than either the unquestionably pernicious heroin or alcohol. Yet still they're prescribed, and still doctors say that they're the only way to satisfy the people who need them to handle their depressions, their fears, and the inability to cope. What gives? A look at the historical record can show how the foul-up came about.

Just to make the idea of an alternate approach clear, we shall veer strongly toward an absolute social reductionism. The Valium sellers do no less with regard to physiological reductionism; somewhere in between undoubtedly is where the real truth lies.

Sociology of Diagnosis

Diseases are not as absolute as they seem. There are fashions in disease, and their hold on how people think can be overwhelming. In antiquity one of the most talked about diseases was leprosy, the horrible skin-destroying infection which could suddenly strike one person, while leaving the rest of his family or neighborhood unscathed. To the pagan Greeks and Romans this was the ideal expression of an all-powerful Destiny. For the Homeric heroes destiny worked by giving them a unique power in battle or leadership; here it worked by picking out an individual and striking him with a disease.

Only in a very few regions was this view not held. The consequences of this in one of those places, ancient Palestine, were immense. For in that one minor province of the Roman Empire the Hebrew tribes living there had developed a system of excellent hygiene, enforced by purification specialists, who at the time were thought of as ritual priests, but whom some modern historians have called health police. The laws they enforced called for such things as washing hands before meals, and not eating potentially contaminated meat, and are at the basis of the kosher laws that many Jews still observe. With these sanitary measures, and also due to the fact that the sun-baked sandy soil of ancient Palestine was a poor breeding ground for dangerous microbes, there were few infectious diseases, including leprosy.

There were, however, other problems. Since the community that enforced these health laws was very tight-knit, there was a lot of stress on anyone who thought they had acted outside the proper customs. This was made even stronger by the belief that a violation of the community's rules was sinful, and contrary to God. The result was likely to be a number of stress disorders, which as many recent studies of repression leading to hysteria have shown, would include lameness, skin disorders, temporary blindness, and the like. Now such a community would be the ideal place for a healer who treated sufferers by compassionate suggestion. And that is exactly what one Jewish, kosher-law-observing young man was recorded as doing.

In Matthew 9:2 Jesus cures a paralytic by saying "your sins are forgiven"; an efficient way of striking directly at the cause of most tension in such a close-knit, theological community. Later on, in Matthew 9:27-31 he is reported curing two blind men through the sensible expedient of acquiring their confidence ("Do you believe that I am able to do this?") and then melodramatically insisting that they really had not stepped outside the rules of their community ("Then he touched their eyes, saying, 'According to your faith be it done to you.' And their eyes were opened.")

For over 1,000 years Jesus' system of healing through touch and suggestion was at the heart of most Western attempts to cure disease. Its efficacy must be reckoned a partial cause of the successful spread of the early Christian Church. Such direct suggestion was quite likely to work well in many areas around the Mediterranean where depopulation after the fall of Rome had produced a low concentration of ambient infectious microbes.

With the increasing population densities following from the improvements in agricultural technique in the Middle Ages, there began an ever-higher incidence of cross-infection, and against that direct suggestion was useless. The worst of these was the Black Death, which was so deadly that in a few years in the mid 1300s it killed something like a third or more of the population of Europe. (French craftsmen who fled from their towns as it approached were given to crying out "misfortune," which in medieval French came out "*plaga*," whence our English label for it, *plague.*)

Plague fit in perfectly with medieval notions of the world, for as opposed to the individually selective leprosy of the heroic Greek days, plague struck down whole masses of people at once, as if by divine chastisement. It was the ideal confirmation of the medieval notion that existence was tragic and man was doomed, and although other diseases probably had a greater effect on mortality in the whole of the Middle Ages, it was always referred to as being the typical disease, for through it divine displeasure with man could be so easily seen. This meant that many other types of ailments, which happened as frequently in the Middle Ages as now, were most unlikely to be thought of as real ailments if they didn't fit in the pattern of plague.

One of the most spectacular of these misrecognized differences occurred to the French peasant girl who became known as Joan of Arc. In 1425, at the age of thirteen, she seems to have had a severe attack of tinnitus, which is a disorder of the auditory nerve from the ear. It produces an unclear noise, which occurs in one ear only, and is often accompanied by vertigo and bright lights on the same side as the noise. These are exactly the symptoms Joan recounted to her English inquisitors six years later, at nineteen, when she was being tried for her life. She said that the first two times the attacks happened she heard an unclear noise, in her right ear only, saw flashes of light on that side, and fell to her knees in dizziness. Now if that happened today an ear, nose and throat specialist would be sent for straight away, but in 1425 there was no thought that ears could make you ill. Illness was what you got from the plague or its like, quite simply, so something that wasn't anything like plague had to be something other than illness.

The result in this case is well known. Joan had a third attack as a girl, in which she dimly made out a few words (which also is standard in tinnitus), and fitting in with medieval theology took them to be the word of heavenly angels. Fitting in with early nationalism too, she took them to be calls to drive the English forces then ruling her country back across the Channel. Young Joan got to work. Only after several spectacular battles, and a crowning of the French king, was she captured by a French turncoat, sold to the British commander, and before him tried, convicted, and burned at the stake. And all, in strong part, for a preconceived lack of a diagnosis.

These conceptual limitations did not get better later. Fashions just changed. In the Italian Renaissance a vast number of ailments were put down as due to syphilis, possibly brought to Europe from the New World by Columbus's crew. But perhaps more importantly, this new disease fit in perfectly with the new image of life being formed during

the Renaissance. It was just right for a society where traveling was increasingly common, promiscuity was admired, and snuggling up to a
king in his boudoir was the best way for a woman to get ahead at
court. It became almost a sign of status to possess syphilis, just as
somebody today who can come into work with a bandage on his leg
where he got a sprain on his skiing holiday gets a certain amount of
social respect.

Once again this popularity, this appropriateness of syphilis, meant
that other diseases just weren't noticed. An extraordinary instance of
this closed-sightedness is the neglect of typhus. This was a terrible infection that regularly killed up to half the soldiers camping outside a
town during one of the sieges that were so common then. But it was
just a bit too much like the now-out-of-style plague, and not at all the
kind of snappy, up-to-date disease that was worth noticing. Because of
this there were few efforts to check typhus, even though whole armies,
and sometimes navies, continued to be regularly wiped out by it. It
was out of date, and ignored. Different social conditions, different disease fashions.

By the early 1800s the particularly unpleasant disease of tuberculosis had priority in cultural attention. It produced the pale look of artistic otherworldliness that was so appreciated among Romantic poets
and those aspiring to be Romantic poets, even though the cause was
not intensity of artistic commitment, but a messy amount of internal
hemorrhaging. It also was an egalitarian disease, fitting in with the increasing democracy of the time, for ordinary city-dwellers in the
highly polluted cities of the new Industrial Age had a fair chance of
getting it too.

In the late twentieth century tuberculosis and its various predecessors have been replaced to some extent by cancer ("That pain in my
stomach, it isn't cancer, is it?"), but above all by what can be termed
medical individualism. This is the belief that whatever illness a patient
has must be somehow due to the patient alone. If a housewife finds
herself moping listlessly through the day it must be because of something wrong with her, the view holds; there's little thought that the
cause might be her empty existence, as an unappreciated housekeeper
and sex object, trapped for hour after hour in the intellectual wastes of
the suburbs. If a college student finds himself crying in the loneliness
of his room, it is conventionally thought to be because he's too weak-
willed to take proper advantage of the campus offerings, and not that
his university is more competitive a setting than any eighteen-year-old
should be thrown into, or that elitist teachers stay safely behind their
lecterns and don't bother to talk individually, after hours, with their
ostensible charges. How much harder it is to think that the problem
might not be in the patient, but in the forbidding setting he's forced
into; how much simpler to give them a pill that dazes their brain cells
enough so that they don't worry about much of anything any longer
(how much scarier for the one be so treated, too).

That's precisely the setting that Valium and crew fit into now. When
the chemicals that transmit signals from one brain cell to another were
first discovered, it was only natural for scientists with this mindset, our

fashion that the patient is the cause of all his problems, to decide that mood and behavior could be changed just by playing around with those chemicals. The reason was that since the person who has depression is the one being blamed for it, then the state of his brain chemicals must be the real cause behind it. The drug companies backed it all the way. In fact it was a staff member of one of the biggest drug companies, the Swiss-based Ciba corporation, who first began using the word *tranquilizer* in the current way in the early 1950s. The implication was that if you wanted to be tranquil, the thing to do was take tranquilizers.

The presupposition that brain chemicals must be the cause of depression or plain low spirits is, alas, only too important for all those 40 million who take the mood drugs. The evidence that depression is caused by the brain chemical changes outlined earlier is very slight. When you take Valium that steps up the chloride leaks inside your brain, true, but that probably only cures low spirits by shaking things up so much that the brain's incapable of even recognizing what a state it's in. Ditto for cutting down dopamine levels in treating more severe cases. Many of the best researchers not working for the drug companies are certain that dopamine or chloride ion imbalance is only one slight aspect of what's really going on in long-term unease or depression and certainly not the sole, independent cause.

The popular drugs treating dispersed, unfocused fear don't really get the underlying brain chemistry right, let alone touch the real cause of an unfortunate life situation. That's perhaps why they need to be prescribed so much, over and over for those 240 million prescriptions. Valium even produces withdrawal symptoms, having such a "sturdy" design that the body stores it away, and takes a long time to get free of it. A look at the history of disease fashions would have shown better.

Let us consider this a sufficient initial surveying of our unlovely topic, and turn now in recompense for our labors to another subject, that is less fear-inducing, and more enticing, by far.

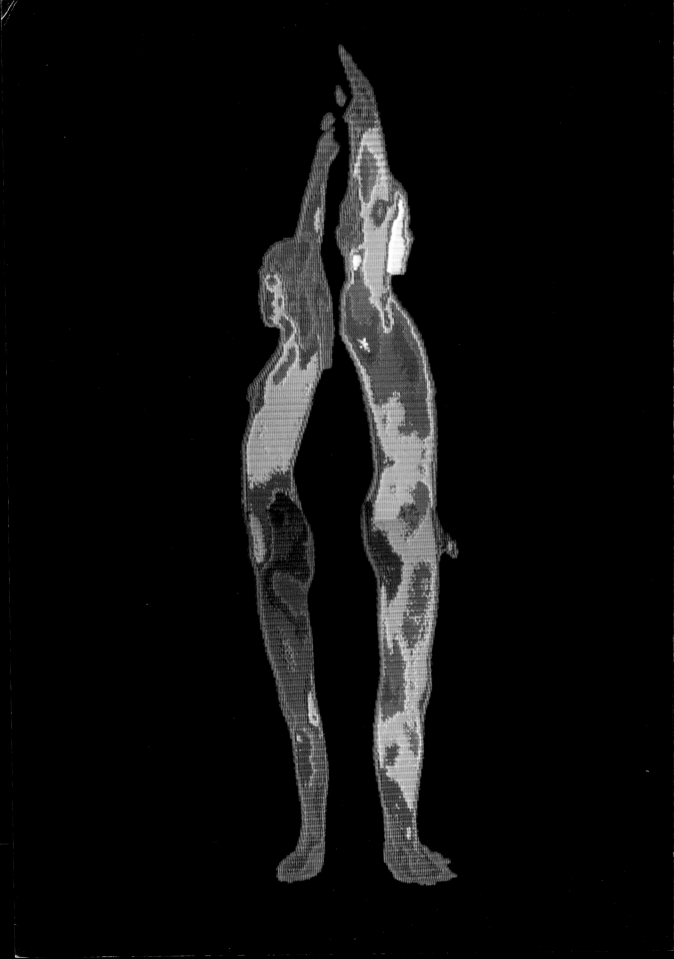

3
Sexual Desire

*I*t starts the same way every time. A mixture of sights, smells, and memories zips along the nerve-cell circuitry throughout the body and ends up swirling into a collection of specially wired nerve cells safely stashed in a ring formation near the bottom side of the brain. This is the limbic system. In animals its stimulation has been noted to lead to faster breathing, faster heartbeat, meaningful gestures with the shoulders and hands, facial grimacing, erections, grooming activity, stomach rumbles, and sometimes impulsive, yet hesitant, aggression. In more refined humans of course, it merely sends off with a rush brain signals that will produce the appropriate feeling, and if conversations and customs go the right way, the desired actions too.

The eyes are always the first giveaway, for the simple reason that they can be considered a part of the brain that has spread forward, and is held from dripping out by the eyeballs. When sexual interest rises, blood pressure starts on a steady path upward too, and the eyes' pupils open wide to help. In less than one fifth of a second they can start to move, growing from a tight 2 mm to a great 8 to 9 mm orb. Tiny muscles in the colored part of your eye, shaped like wagon-wheel spokes and reaching into the pupils, are what do the trick. (Other strong emotions can make this happen too, albeit not so much, which is why Chinese jade dealers have long worn dark glasses during their bidding meetings to hide the clues of their real interest that their swelling pupils might otherwise provide.)

With pupil dilation comes the almost reflexive move of aiming those now extrasensitive eyes all around whatever object may have been the original source of arousal. It happens in great ballistic jumps

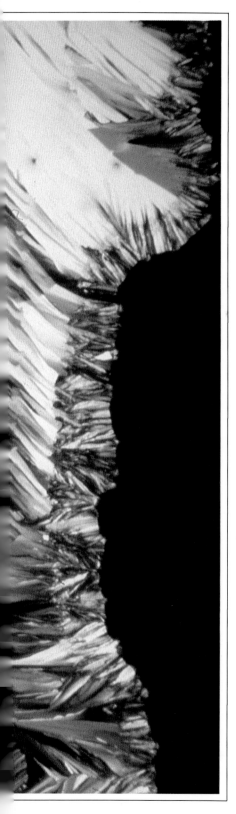

of the eyeball, as six toughened muscles in the eye socket tug the eye-ball from one position to another, at speeds far greater than the fastest twist the lumbering whole head could manage. The normal blinking rate of one sweeping closed every six or so seconds is likely to be cut way down when a highly eager pupil is concentrating. Without further help that would leave the easily drying eyeball open to dust, wind, motes, cinders and cracking. This undesirable fate is neatly avoided by an automatic mechanism that pumps down on the little tear-containing sacs wedged beneath the skin just behind the limit of the eyes, send-ing out from these sacs a minuscule spray of tears, rather like washer fluid arcing its way onto the windshield of a car.

This minute squirt and splash on the surface of the eye in times of imminent passion is figuratively seen as references to the dewy-eyed gaze of eager lovers. The enlargement of the pupils is less readily seen, though most sexually experienced adults seem to have enjoyed at least a subliminal effect from its invariable occurrence. In tests where a number of men are shown two photos of a woman that are identical but for the fact that the second has been touched up to en-large the pupils, almost all the men will prefer the second picture, usually without being able to specify why. This phenomenon was rec-ognized by noblewomen of the Italian Renaissance, who would fre-quently drop an extract of the deadly nightshade plant, belladonna, onto their eyes to produce the enlarged pupils that might be espe-cially attractive to their suitors. They did not do this just to raise their passive appeal. The investigators who first did the experiment on showing men the different photographs found that when a man said he preferred the second picture, the one of the woman with the en-larged pupils, then *his* pupils enlarged, too. This response to the re-sponse produces an ultrarefined body language, and one most easily followed by a participant with just the extra light-gathering abilities produced by the belladonna of the apparently fawning Renaissance women.

Hormones

Changes in the eye are only the quickest sign of sexual interest. For more long-term arousal there are the aptly named sex hormones. A small amount is always swirling around in us. Behind every quick arousal by a suitably enticing memory, fantasy, or real-life companion,

Testosterone, the immensely powerful male sex hormone. Produced in the testicles from puberty on, it initiates then maintains beard growth, shoul-der proportions, penis length and sperm production. Without regular production of less than a single drop secreted into the bloodstream, all these changes would gradually cease.

a whole convoluted chain of hormone-associated events will need to have taken place.

The hormones begin in a tiny section of gray-colored cells wedged in the middle underside of the brain, the hypothalamus (Greek for "under the inner room," neatly describing the shape of the structure over it, and its bottom most position relative to it). In that area, somewhat smaller than an average prune, a small storehouse of a triggering chemical called LHRH which has been patiently waiting, gets jammed into a series of narrow, downward-stretching channels, tumbling a half-inch down until finally, the pituitary gland has been reached. This is the very lowest limit of the brain, dangling from beneath it at about the level of your sinuses and being not much larger than a pea. It doesn't have much space to store this chemical that is squeezed down from the hypothalamus overhead. Instead of keeping the LHRH to itself, it pushes another chemical along the narrow blood vessels that run conveniently around its edge. Plopped neatly against these blood vessels, this hormone gets sucked on into these, and, once within, starts on a buffeting ride that takes it all through the bloodstream. The two-step action starts automatically once every hour, and goes on, in these chambers and tubes back behind the nose, hour after hour for all of your adult life.

The second chemical has little choice in where it goes. Carrying its sexual secret, it rides along with the bloodstream as it swirls down the neck, trickles across the lungs, and roars past the elbows. It finds not the slightest call to stop its journey until it sloshes into the vicinity of the only cells that will cause it to stop and grab hold: the female ovaries or the male testicles.

For simplicity, let's take the latter case. When the triggering chemical has been swept to the testicles, it locks into certain of the cells there, and by its very presence makes them let loose with all of the testosterone they have on store, and then to go on making more. This testosterone is the main male sex hormone, and is made from cholesterol of all things, as the prime ingredient. ("Testosterone" specifies just a chemical from the testicles. The equivalent female sex hormone has the more melodramatic name of "estrogen," taken from the Greek for "producing frenzy," and showing the biases of the male anatomists who thought the name up.) Once the testosterone gets released into the thousands of blood vessels and lymphatic tubes that wend and stretch all around inside the testicles, it sweeps through the body in its own right, raising general sexual interest and helping ready tubes, tissues, and organs that might be called upon later.

These sex hormones are always with us and are much desired, for if for any reason they were to be totally removed, women would find their breasts, labia, and clitorises all would suddenly tend to shrink, while men would gradually lose their sex drive and see their organs slowly begin to shrivel. Even a mild lowering of certain sex hormones will stop the libido in both sexes (through raising the hormone seems to only raise libido in women).

The hormones are even what make an apparently sexless embryo into a noticeably male or female baby. If the fetus is going to be a girl,

*The sex hormones are so power-
ful that through their effect on
general metabolism they can
stop main bone growth throughout body. This gamma-ray scin-
tigram of a normal child with
active growth zones at the ends
of long bones (knee, hip, shoulder), as usual. These will fade
and stop, producing final adult
height, as sex hormone production increases then stabilizes
after puberty.*

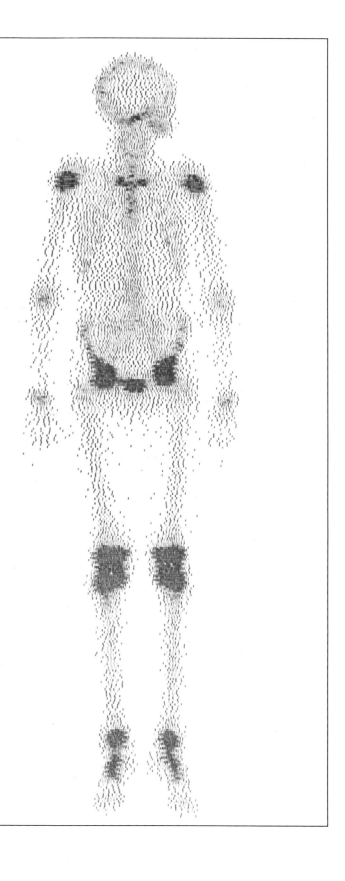

for example, female sex hormones will be swirling around by perhaps the seventh week in the womb, causing the smooth skin at the base of the abdomen to develop into the labia lips and clitoris; in a male fetus testosterone will cause the same tissue to shape itself into the scrotum and penis. These background sex hormones don't have much to do after that until the years of puberty. Then they really come into their own. It's the production of the appropriate sex hormones in the ten- or fourteen-year-old child, stimulated as always by the messenger chemical sent down from the pituitary, that causes all the well-known changes of those years: the growth of the sex organs as well as such secondary changes such as breasts and rounded hips in what were once unselfconsciously streamlined girls, or moustaches and deep-ened voices in boys. Both male and female sex hormones have the side effect of getting the cells at the ends of the body's long bones "gripped" so that all bone growth slowly slides to a halt. This ensures that once a teenager becomes a potential parent, he or she will no longer be burdened by the energy-sapping demands of a still-height-ening frame. That's why the first kids to go through puberty in a school class might be temporarily the tallest of their age, but are likely to lose this distinction as their already fixed bones are outmatched by the late sprouters who perhaps with some glee outgrow them in turn.

In adult life the sex hormones are steadily produced, ensuring that changes from puberty stay there. This has the indirect effect of making humans into perhaps the most sexually arousable species there is. A contrast with the animal world would help explain why. Most large mammals are sexually receptive only at a time that will ensure that any offspring they produce will be born in the spring, when the weather's not too harsh and food's there for the taking. For most species of sheep and goats, with a six-month pregnancy period, only in the au-tumn is there any interest in mating, for only then will conception produce a springtime offspring. For wild horses with a one-year preg-nancy, signs of sexual interest and mating are restricted to a brief pe-riod in the spring, ensuring birth the following spring, while for the noble elephant, which goes through a whopping twenty-two-month-long pregnancy before a three-hundred-pound, hairless infant emerges, sexual interest is restricted to the early summer two years before the birth.

That makes humans the odd ones out. By all rights, courtship and consummation should be confined to a few weeks in August. Breasts should pop out and sexy voices should come on strong in those weeks, only to pop back for the yawning boredom of the rest of the year once the necessary work has been done. The reason we are spared such disconcerting changes is that sex hormones make no such sudden changes. At puberty they turn on the sexual signs, as well as the sexual powers, and hold them in the "on" position for decades and decades to come. Humans are so much interested in advertising this full-time availability that artificial additions to the work of the hor-mones are devised. Medieval noblemen would often strut around with great bulky lumps of cotton over their ostensibly private parts, so to suggest the size and interest in what was at the sources of this swelling underneath. Not long after that went out of style, the ostensibly more refined replacement fashion of neckties came in — though the fact that they were worn exclusively by men, were long and straight, and dropped down the centerline of the body, left little doubt about their symbolic purpose. An older man with a very wide tie gets glances for assuming too much, while the cruelly truncated version called the bow tie, goes together in the popular imagination with college professors and other wispy types. Women have gone through extraordinary lengths through the ages to emphasize the slight waist/hip curve that their sex hormones indirectly produce, while the moistly inviting circular redness of the facial lips, produced by a thinning of the skin over the red-tinted capillaries there, is considered so necessary to enhance, that lipstick sales hold steady even in a widespread recession. All these artificial aids are just attempts, conscious or not, to enhance the full-time signals that the constantly swirling sex hormones help produce.

Sometimes, the swirling is not just right. Sometimes a fetus genetically programmed to become a girl will receive an unexpected hit of its mother's sex hormones, a hit normally avoided by the placenta between mother and fetus, which is precisely designed to prevent such direct mixing of blood. These intruding hormones will stop the normal development of female anatomy, and the fetus is likely to be born with enough of a male's genital shape to be raised as a boy by the unsuspecting parents. Only at puberty, with the uprush of female sex hormones within the child, does the trouble begin. The result is likely to be a youngster with an external anatomy that's neither fully male, having but a poorly defined swelling where the penis should be, nor quite female, having something looking like labial lips but nothing more than some folded skin where the vagina should be.

Such unfortunates represent an extreme case of a situation we all share: for all women have a good 5 percent of their sex hormones made up of the "male" testosterone, while for men just as much of their sex hormones is the ostensibly "female" estrogen. For some unknown reason this is the mixture at which our systems work best.

Male Anatomy

What happens next, often not without considerable trepidation, is up to the organs themselves. The key role of the texticles in these proceedings was recognized by the ancient Latins, who gave them their word *testis* for witness; their existence of a healthy pair under the toga being a witness to the bearer's capacities. "Testify" and "testimony" both come from the same root. More lighthearted in descriptive terms were the early Greeks, who used the word *orchis* to mean testicles, and upon noting that a certain common flower came from a root of that distinctive semispherical shape, thought it only just to call it by the same name, which is why we call an orchid an orhcid. If the Romans had been consistent and renamed the same plant in their preferred nomenclature, then all eager young females attending their senior prom today would be only too proud to sport a luxuriant testicle pinned neatly onto the upper corner of their dress.

The testicles' apparently precarious dangling position is due to the need to keep them cooler than the rest of the body. The cells in them that get acted on by the messenger from the pituitary and produce testosterone could possibly get along fine at the higher temperature inside the body, but the other cells in them which serve to produce sperm cells certainly could not. When the temperature goes up the sperm count goes down. A clever heat exchange system is placed to run hot arteries past cooler veins, but it only helps a bit.

Many fertility specialists have been able to turn anxious patients into happy fathers by simply telling them to change their daring tight nylon underpants for sensible loose cotton ones. The outer covering of the scrotum is just a bit of loose walling that back in the early embryo stage stretched around the intestines, but later slipped downwards. It has plenty of small muscles on call to move it up or down as needed.

Most notably, the outsides of the testicles are wrinkly. That happens not because the skin doesn't fit well enough, but because it actually fits too well. In most parts of the body, wherever the outermost layers of your skin fit over the deeper, supporting layer, there are wedges, cords, and splints linking the two. There are only two significant exceptions. One is the eyelid, where the function layer is smooth, the other exception is the outside of the testicles, sliding and easily wrinkling about the same reason.

Since cooling is the main reason for the external placement, any position outside the body will do for the testicles. Hanging from the ears would probably be ideal as air flow goes. Some animals have made a start. Certain creatures have them hanging up on the belly almost, in *front* of the penis, while some gibbons among the trees swing with their testicles blithely off to the side of the penis. The aft position in humans provides perhaps a little more protection, but not of course a great deal.

Societies throughout the ages have found a use for such disburdened individuals. The very word for a castrated male, "eunuch" shows one. It comes from the Greek for "guardian of the bed chamber," for it was widely held that a man without testicles could be safely employed in guarding a ruler's harem. But the eunuch's lot was rarely the unmitigated woe outsiders thought. For only a boy castrated before puberty is unable to overindulge in sex; someone castrated after that will have enough testosterone produced by outer layers, or cortex, of the multipurpose adrenal gland as an automatic fallback to be able to have erections and sometimes even to ejaculate. Only the 200 to 400 million sperm cells which the average ejaculation has would be missing, but as they make up less than 1 percent of the semen by volume it would be unlikely to be commented on. Only in the imperial Chinese court was this indulgence definitely cut out. Candidates for Imperial Eunuch-hood were dealt with by a special, sickle-shaped knife that removed both the testicles *and* the penis with a single swipe. Subsequent bleeding was usually fatal, for there's a big blood flow to the region (it shines clearly in thermal scans). Yet there was never a shortage of candidates, usually put forward by hopeful parents who hoped to raise their families' standing.

Some of the 200 coiled sperm-producing tubes, that give the testicles their distinctive texture. The scattered flaky sheets in front of them, that look hardly useful, actually produce sex hormone testosterone; contain bacteria-eating macrophages; release histamine to regulate blood vessel width; and lead blood and lymph passageways into production tubes to supply sugar, fats, oxygen and hormones. ▶

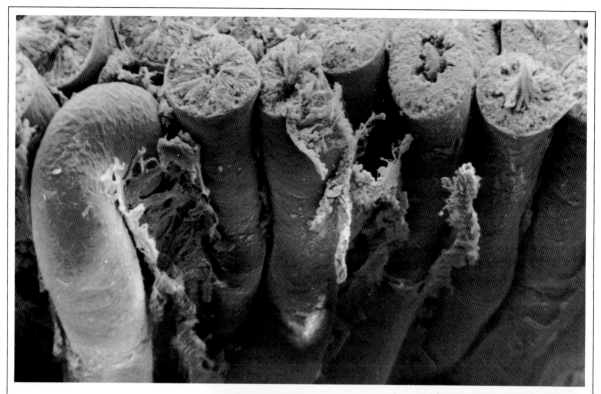

Details of sperm production within the tubules revealed at higher power. Rounded cells just inside the circular rim produce simple copies, which develop in several days' journey to center of tube into mature sperm. Their tails can be seen as the tangled white wad in center. Wedged between the developing sperm are other cells that secrete chemicals which pull in testosterone, thus providing local concentrations of that hormone high enough to help sperm complete their taxing growth. Total production rate: 500,000 completed sperm per minute — each with an exact cloned copy of possessor's genetic information. To give an idea of scale, a man reduced by magnification of this electron micrograph would be no taller than the thickness of a toenail. ▶

The few who survived the ordeal were thought to provide the ideal disinterested advisors to the emperor, for unable to found a dynasty themselves, they would have no motivation to give advice that could lead to his overthrow (except, of course, revenge). Power politics was not the only reason for the dead. The masterpieces of eighteenth-century Italian opera almost all achieved their delicate soprano highs with choruses of young, and not so young, men composed almost entirely of what were gently referred to as *castrati*.

Luckily, such disencumbering is rare. In properly attached testicles sperm production goes on constantly in over 600 tightly wrapped tubes, so slender that if unwrapped and twirled around they would make a lasso over half a mile long. In that generous space some 200,000 sperm are produced per minute. That's 3,000 a second, every second. The ongoing rush continues while paying for a film ticket, switching on the turn signal in a car, or stepping into a crowded room. Squish, slither, plop, and there they are.

When these mammoth numbers were first discovered it was thought that increasing buildup must be the source of the generalized sexual desire that arises after any long period of abstinence. But no such pressure receptors were found, and the theory has had to be scrapped. After all, the buildup is more impressive in volume than in number, for each individual sperm cell is less than five hundredths of an inch long and a hundredth that wide, so it would take a solid phalanx ten thousand in a row and several hundred in a column to appear as even a tiny dot. Unused mature sperm is dissolved and then absorbed by the body after a few weeks.

Each of the two testes is surrounded by a snug, tuniclike covering, which is just loose enough to give each one a buffering space it can slip around freely in without getting damaged by friction or pressure. Also in the loose connecting network around the tubes where the sperm get formed are a whole population of uninvited kibitzing visitors. There are squashed, cigar-shaped cells, free-moving bacteria-chomping cells, cells that store little balls of the inflaming chemical histamine, and many, many, more. What's called the "blood-testis barrier" is set up in the testicles to keep the sperm in their little tubes safely away from everything else without.

The average sperm has a smoothly oval head placed on a long, tapering tail, but, as usual, the average is not the rule. All men produce a mix of wildly different sperm, with each being specific enough for Scotland Yard's analysis laboratory to have developed an expertise in taking "sperm prints" analogous to fingerprints, though not so precisely incriminating. Since the prostatic fluid in which sperm are bathed is a function of the male's blood group, better detecting, or at least exclusion, can be done that way too. Milling about with the standard sperm are always a mix of other ones. They'll have two normal heads, or a round head, or a flattened head, or a bloated head, or two normal tails, or an overwide tail, or some other distinguishing oddities. This is the hodgepodge from which life ultimately emerges.

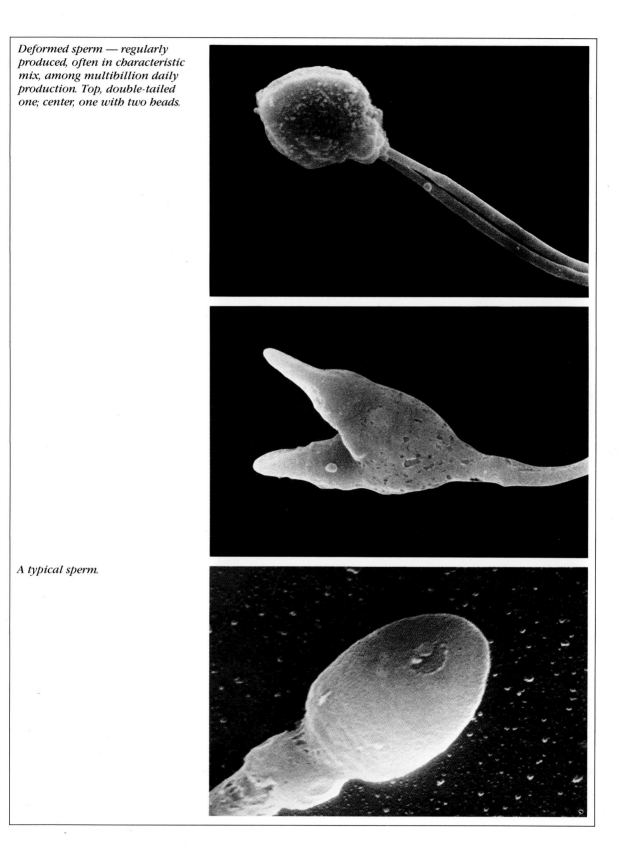

Deformed sperm — regularly produced, often in characteristic mix, among multibillion daily production. Top, double-tailed one; center, one with two heads.

A typical sperm.

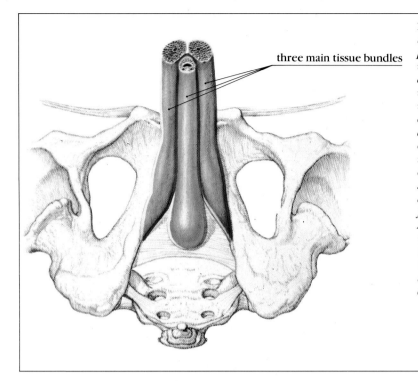

three main tissue bundles

The three main tissue bundles of the penis. The two outer ones are paired, forming a deep groove in which the inner one fits. All three are composed of a dense lattice-work of open spaces, which are what well up with blood during erection when outgoing drain-age veins are blocked and in-coming arteries unchecked. The middle one contains the urethra channel through which urine or semen is transported; it widens at the end (not pictured here) to form the nerve-rich glans penis. All three columns extend from well inside the abdominal wall, under the pubic bone; the chang-ing depth of emplacement with age can affect the angle of erection.

As the ungainly mix of sperm emerge from their production tubes, they enter upon a weird landscape, still within the testis, of 5 to 10 un-gainly ziggerat-shaped structures. These have little waving hairs inside that ripple the still baby sperm along through them. Then there's a rest and more growth in the main tube leading out of the testicles.

After that, the sperm have gaping before them a two-foot tunnel that goes for an amazing meander in the lower abdomen. At first it rises near the surface of the scrotum, then passes up and into the abdomen along the side of the pelvic wall, then stretches behind the bladder, and then returns toward the front again at the base of the prostate gland. Whew. But there's still more, for at this point it junctions with the exit tube from the bladder, and there is a delicate valve system to ensure that semen and urine never pass beyond at the same time. (It takes a while to switch from one routing to the other, which is why it's not possible for a man to urinate during an erection, or for a brief time after ejaculation.)

From that junction the sperm's exit tunnel plunges into and through the penis. This exit fount is composed of three long slivers of erectile tissues — two are paired and form a deep groove, while the third lies like a toffee column right between them both. Each has a giant 3-D network inside perfectly repeating nodular spores that can soak up blood. Under the right signals the arteries feeding into these three tissues will pump in more blood than usual, while the veins which usually serve as an exit conduit for the blood get squeezed tight by little muscles around them. (Whales, curiously enough, have similar foam-filled sinuses around each ear; swelling up quickly they provide an excellent protector against changes in pressure for these deep-diving creatures.) This crucial moment of opening the sluice gates is further helped by a powerful artery-opening chemical, called VIP, that was originally discovered in the intestines, and that here is stored up in tiny bubbles all throughout the base of the penis and two of the three erectile tissues. There are thousands of nerve tendrils with nothing better to do in life than wait there in the penis, the VIP bubbles ready inside them, and pop them loose when the erection call from the brain begins. With more blood going in than coming out the penis swells up in what can be as little as 3 to 5 seconds, to a length which averages 6¼ inches long, and 4½ inches around.

The precision of those figures is sadly symptomatic of the truly obsessive concern with which penile changes are so often treated. In most areas of the body rounding off to the nearest inch would be accepted with no problem, but here the experimental subjects must have held out for that extra quarter of an inch.

Such proud fractions pale when contrasted with the 18-inch erect penis of a boar, the 30 inches of the stallion, and the seven-to-eight foot penis of that largest of all creature, the blue whale. The blue whale comes in for special note, for while many animals get added support from a bone stretching the length of the penis, something found in the bear, gorilla and lion, the blue whale makes do with just the might of its veins and arteries, aided, it is true, by the natural buoyancy of the sea.

Female Anatomy

In women, the key point is where the triggering chemical from the pituitary goes, in this case the two ovaries. These are wrinkled and flattened, bean-sized, if not smaller, and rest on muscular bits that stick out from either side of the abdominal wall. Inside each ovary at birth are over 300,000 minute egg cells, as small to the ovary as the ovary is to the World Trade Center. They don't do much of anything until, with puberty, they start to emerge, one at a time at an average of one every four weeks. That translates as 400 eggs released in the whole 35 years or so of fertility, a number smaller than the quantity of sperm cells a man's testes are producing every half-second. Here quality wins over quantity though, for the full-grown egg cells are hundreds of times larger than the sperms, filled as they are with all the supplies needed for the first few days' feeding of a fertilized embryo.

The egg that's going to come out in a given month starts as a tiny speck wedged firmly in the center of the ovary. Slowly it squeezes its way to the outside, growing wider and rounder as it moves to the surface. Once there, it pops loose, leaving a small crater which for several weeks will be the sources of hormones readying the body for pregnancy. The popping out and release of blood can hurt and is the source of what can be the startlingly sudden pain of ovulation. The crater gradually evens up, but never quite goes away. A few decades of that and the ovaries of a forty-year-old will look uneven and pitted, forming a gnarled contrast with the smooth ellipsoid of a fourteen-year-old.

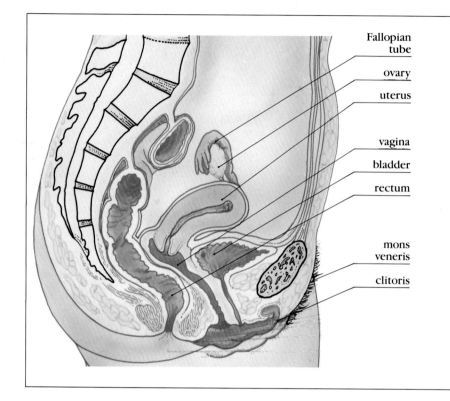

As we will see at more length in the next chapter, the loose egg is not loose for long. Stretching wide, full over the top of the ovary, is the trumpet-flaring outer entrance of the fallopian tube. Into this passage the lone egg is pulled, swept along halfway to the uterus by undulating muscles and cilia inside the tube. There it waits, possibly to be fertilized by a blindly stumbling sperm cell, possibly to arrive untouched in the uterus and be sloughed off two weeks later during menstruation. Between this egg, waiting in its fallopian tubes, and the outside world there are two more main areas. The first is the uterus, a muscular hollow normally about the size of a fist, while beyond that is the outermost of them all, the vagina. Both were first analyzed by male anatomists, and many of their preoccupations linger.

As one example, the men of ancient Athens believed that women's changing moods were caused by their uteruses wandering inside their bodies, often ending up lodged just under the neck. Accordingly, their solution for female depression was to have a female slave remove the chiton of the offending sufferer and place a warmed container of spiced incense close enough to the vagina so that its attractive vapors would ascend into the body and lure the recalcitrant uterus back down to where it belonged. The underlying idea was that while a man's moods must be due to the influence of the gods, a woman's problems were due to her unfortunate anatomy. As uterus in Greek was *hystera* this gave us the two usual pejorative variations "hysteria" and "hysteric," terms usually applied to women which have remained with us to this day.

The Romans were of an equally unfair bent. Roman worthies of the Imperial Age often went around their business with clanking swords held on one side of their togas in scabbards, a receptacle which in the Latin of the time was called a *vagina*. Mighty war leaders like Caesar were undoubtedly proud to have an exceptionally large vagina, or even a pair of them. This chauvinist culture labeled the passage from the uterus a "vagina" too, codifying the notion that it was significant solely for its abilities to serve as a sheath for a man's penis, and not because of the pleasure it provided for women.

But pleasure it does provide, aided closely by the various tissues scattered strategically at its base. They include the outer folds of the skin and fat called the labia majora, or outer lips, and the thinner pair within them (due to lack of fat) called the labia minora, or inner lips. The analogous names are misleading. The outer lips are simply formed of ordinary skin. They fuse together on top, just where there is a stocked-up pile of fat and connective tissue where the pubic bones have themselves fused to each other: the mons veneris. The inner ones though are lined with the quite different substance called mucous membrane, the surface that is found on the tongue, the throat, and the inside of the eye. Nor are they really the innermost lips. Tucked away as a further symmetrical entrance within *them* are the grandly named bulbs of the vestibule, not bulbs really but two thin strands of sensitive tissue. To complete the genital geography, just above the vagina where the inner lips join is the sensitive tissue of the clitoris.

All of them undergo changes during arousal. The outer lips spread wider than usual and then curve out flat, while the inner ones enlarge and stretch straight forward. The exposed clitoris gets longer, often doubling its length to something over a third of an inch — though the variation here can be tremendous, with up to three inches being reported in some. The upper part of the vagina starts to swell, in something of an erection, as the muscular bands holding it in place relax. It also begins to get wet with four separate processes taking place: swelling blood vessels all the length of the vagina push moisture beads through its relatively thin walls, when the beads get dense enough they suddenly merge and form a thin but continuous slippery sheet. A special fluid-containing gland near the entrance trickles out some of its slippery contents, and finally somewhat thicker secretions come out higher, at the entrance to the uterus itself. All this lubricating mucus is composed of a very simple protein, called mucin, mixed in with a special sugary acid. When produced in the mouth this mucin is a central ingredient of saliva. Here in the vagina, the loose acids about force the tiny mucin pieces into an interlinked, somewhat log-cabin-like structure. That internal bracing is what gives the mucus here its exceptional viscous properties.

A change in mood can end matters here, but if not, the changes continue. The clitoris tightens and often gets pulled back out from under its usual hood by two bundles of muscles that hold it in place, and stretch down on to the pelvic bone. The outer third of the vagina begins to contract, which can be useful in gripping a penis, or can be

Progesterone, one of the chief female sex hormones. Secreted from the open crater left on the ovary surfaces by a released egg, it works with other hormones to prepare the uterus for pregnancy. Potent in other changes too, it is so powerful that as a partial cause of premenstrual tension it can cause the body to retain over six pounds of water — ⅔ of a gallon.

delightful on its own. The outer lips might swell a bit more, but they are definitely upstaged by the inner ones. These undergo a tremendous color change, from dull pink to bright red. In the case of women who have had children, they keep on going down the spectrum till a burgundy color is reached. The sheer volume of blood in the body increases. Stopping at this point can be excruciating, yet it is necessary to hold matters in abeyance for a moment to examine why.

Biological Reasons for Sexual Arousal

Why have sex? The immediate answer is because it feels so good, but that just puts the question one further level along. Why *are* we so constructed that a pause at certain delicate moments is the last thing desired, and that in hope of such moments people will go through extraordinary rituals in the course of the day? The proper Victorians had a simple explanation for this, which was that sex was necessary for the continuation of the species. But many species get by fine with solitary or asexual reproduction. One such is the amoeba, which just divides in two when more amoebae are called for. (Many of the multipronged bacteria which live in great colonies on your face do the same.) This is fine if all that are needed are identical amoebae, and in the amoeba's preferred home of quietly stagnant water, such reproduction will usually suffice. Higher creatures however originated in environments where more exciting things were happening.

Only with sexual reproduction, in which two individuals combine their genetic inheritance so as to produce a third one who will be somewhat different from either of the originals, can such a changed environment be dealt with.

Sometimes it's an improvement, as was undoubtedly in the mind of Isadora Duncan when she proposed to George Bernard Shaw that they have a child on the grounds that it would have her looks and his brains. Sometimes it's not an improvement, as was realized by Shaw when he declined the offer, observing that the child might well end up with his looks and her brains. Usually it's neither one nor the other, but just a blended form that is unlikely to do much better than either parent if circumstances stay much the same, but at least has a fair chance of doing better than them if the world it inhabits does undergo a change. Such changes were frequent enough in the past to wipe out all large creatures that were just clonelike imitations of a single parent. The advantages of mixing genetic material were so great that special incentives were developed to ensure that it took place — including the pleasure of sex.

Sexual reproduction is also needed because the earth is continually bombarded with high-speed particles hurtling down from outer space. While you are reading this sentence, for example, billions of low-energy atomic nuclei blasted out from the sun will pass through every square inch of you at over a million miles per hour. These have little effect, but not so the much smaller number of cosmic rays, which probably come arcing in from right across the galaxy, and whomp into you at enormously greater speed and energy. They can barrel right into a delicately placed atom in the genetic storehouse of DNA in the sperm cells of the testicles, or the egg cells of the ovaries. Enzymes around the DNA frantically try to repair the damage — tugging and stitching the damaged DNA back into place — but as 8,000 cosmic rays can hit your body in just one of these frequent, invisible showers, that repair doesn't always work. There's about a one in five-billion chance of such a mutation happening to any single gene in a generation, (what's called the spontaneous codon mutation rate), and while other factors than just radiation are involved, over time that can add up.

If humans just sloughed off carbon copies of themselves, any such genetic disfigurement would likely show itself in the next generation, and that could be deadly. With sexual reproduction, however, even if the DNA in a sperm cell got the cosmic ray zapping, it's unlikely that the same atom in the DNA of the egg cell it fertilizes will have been affected by a cosmic ray, too. And with only one such fault, the resultant child will often show no disfigurement at all.

Only incest breaks up this calculation, and that's why it's almost universally banned. It's not at all a matter of superstition, for brothers and sisters do share 50 percent of their genes, and there must have been a lot of times when their offspring were terribly affected by the identical genetic faults they both carried. Even in the first-cousin marriages, where the shared genetic material is down to 12.5 percent, severe mental retardation is some three times the usual rate, while in brother-sister offspring it's *twenty-seven* times higher than usual. Bad news.

A Proviso

At this point the biologist must admit a limitation. Other animals are more taken over by their physiology than we are. In certain types of praying mantis the male will continue to copulate with the female even when she leans forward and begins to eat him, starting at the head and moving downward. That's determination. Humans are less mechanical about it all, easily distracted. So in what follows it must be remembered that the explanations of physiology say what can be, and not always what is.

Men (continued)

It is now time to return to the male of the species, last seen in a precarious state of enlargement. The hydraulics of the erection are simple matter of more blood's being pumped in than out, with the spongy tissue in the penis relaxing to take it up. The tiny muscles that squeeze down on the veins coming out of the penis, producing this distension, are moved to their deed by a signal from one of the pelvic nerves. This particular nerve stretches out from the lowest part of the spinal cord and ends in the muscle around the crucial veins.

Any number of things can cause the nerves to stop sending messages to the muscles — anxiety, hesitation, qualms or squeamishness will do. The chemical that the nerve releases, acetylcholine, is like most neurotransmitters in not being a very stable molecule, and after holding the tiny muscles there tight for a while will quickly break down. More signals must be sent along the same nerve if the muscles

Praying mantis — possessor of illuminating copulatory habit (see text).

are to be kept clamped down and the barrier it forms against the high pressure of blood contained on the other side is to be maintained.

When a man finds his erection dwindling away before warranted, or not arising in the first place, it's usually because other centers in the brain have countermanded this repeated nerve-firing and acetylcholine release. That's why conscious concern is not recommended. It only tangles up the normally straightfoward main nerve path. The powerful VIP bubbles gurgling loose in the penis to hold the erection arteries open can also be stopped by too much macho anguish.

The nerves controlling the veins do not have to shoulder all the burden themselves. There's aid from the outermost quarter of the penis, called the *glans,* Latin for "acorn," because of its shape. It's packed full of a tremendous number of pressure receptors and other nerve endings, all laced just under the surface. These endings are very close together, which is what makes the sensation there as fine as it is.

Not every place is so honored. The middle back, for example, has its nerve endings only sparsely scattered. Most people detect two simultaneous touches there as being separate only if they are least 2½ inches apart. On the forearm it's 1½ inches, while on the fingertips, where there are a lot more receptors, it's a tenth of an inch, and on the tip of the tongue it's .025 inch. (That's why even a tiny paper cut on the tongue seems to open a chasm.) The erect glans does almost as well.

It would be a mark against technological civilization if it did not try to extend its efforts into even this area of human endeavor. Men hesitant to stake their all on a single pelvic nerve and the quickly dissolved acetylcholine molecules can get "miniaturized erectile-maintenance units" installed. Early versions depended on hasty pumping of a small bulb at the base of the penis that pushed a hydraulic substance into two or three enlargeable cylinders that had been implanted in the waiting member. The latest version is electric and triggered by a tiny button tucked in line with the skin, which reveals its functioning only by a slight whirring akin to a loud quartz watch. Or so the manufacturers proclaim.

As a normal erection continues, other areas of the body get affected too, as raised blood pressure, pulse rate and other nerve firing starts adding their relatively useless contributions to the act. The results can be slightly distended earlobes, fingertips, nipples and nose. Skin temperature rises, breathing speeds, swallowing slows and apparently for certain unfortunates, contact lenses will pop out.

These changes are significant more as a marker of what is to come, just when the acetylcholine-emitting nerves really come into action, the body gets ready to turn them all off and to switch action to that other set of autonomic nerves, the ones that fire adrenaline or norepinephrine at the target tissues. It is this switch that heralds ejaculation. When stimulation has continued long enough (and that can vary from a measured 2 seconds to 15 hours, the latter in a case reported by Mae West of "a man called Ted" she knew, with the American average being around 8 minutes), this second set of nerves quickly delivers a carefully arranged cascade of signals. Each strikes in sequence on the glands and muscles around the passageway that leads up from the testicles and around and then out through the penis. Once they start there's no stopping the process. It is a most peculiar way in which this second set of nerves works. They start from the brain, run down the spinal cord, and then, instead of heading straight to the interested sectors, the nerves bunch up and race around each other in big, convoluted knots (the ganglia from chapter 2), that are evenly spaced just under the strong muscles of the back, a few inches away from the spinal cord. Only then do they head to the genitals. The orders for orgasm first pass through a staging point in the lower back.

This final wave of signals squeezes down on tiny muscles that send the sperm and additional fluids hurtling up and around to a very temporary waiting depot about halfway through the whole outlet tube, at a

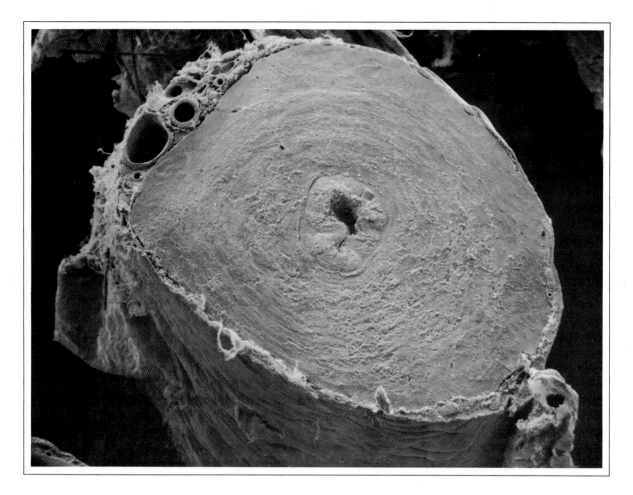

This is what gives the male ejaculation its velocity — the extremely thick layered muscles wrapping the ductus deferens, leading to the final exit path the sperm take from the body. The sperm is propelled down the narrow central channel by a muscular wall pulsing in rhythmic waves, controlled by a noradrenaline release of the sympathetic branch of the autonomic nervous system leading from a ganglion switching station associated with the pelvic nerve. The small cavities at the upper left are blood vessels, which bring nutrients to the muscle.

point where it passes through the prostate gland. After less than a fifth-of-a-second pause, more contractions of the muscles wrapped around the path squeeze in a perfectly timed rippling wave behind the semen, sending it all the way through the exit of the penis. An electro-encephalograph now will show clusters of millions of nerve cells low in the brain's limbic system firing in unison, just as in a mini–epileptic fit. The same instrument shows a near blank-out of cell firing in the rational cognitive centers of the brain's cortex; the whole is preceded by a blast of activity in the vision centers at the very back of the skull, perhaps so explaining the overpowering flash of light often reported at the start of orgasm.

Three or four quick semen bursts are usual, at a regular space of 0.8 seconds apart. Most of the sperm are in the first one. Unobstructed ejaculations have apparently been measured at a length of 3 feet, though 7 to 8 inches is more typical. The muscles lining the urethra, the exit channel within the penis, did their pressing under full control of the muscles of the groin and anus, though this operates more through raising the general pressure of all fluids in the region than by a specific propelling of what comes out. Ninety percent of semen is plain water. Less than 2 percent of the total comes from the testicles,

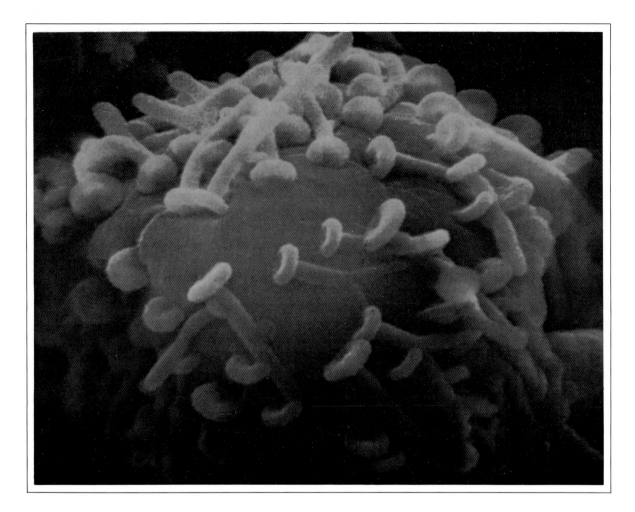

including the sperm cells at the center of it all. The rest comes mostly from the small bulbs of the seminal vesicles, which feed into the outlet channel early, and provide most of the gooeyness — as well as sugary fructose to provide energy for the swimming sperm. (The chemical from the vesicles that makes semen sticky is also to be found in the vitreous humor of the eye, behind the lens, making it viscous too. The aqueous humor in front of the lens, that you are looking through now, is watery and transparent because it has much lower concentrations of this semen-found substance.) A bit over a third of the semen comes from the prostate, providing the characteristic odor. As for semen's color, it's due to oxygen-eating flavoproteins that carry metal atoms in their centers, and which when reflected give a cloudy, whitish hue.

Some 3.5 milliliters of semen, a small teaspoon's worth, will come out in an average ejaculation, though there can be as little as a flying speck of .2 ml, or as much as a globular stream of 13 ml after long abstinence. The caloric content of the average ejaculation, rich in proteins, low in fats, probably measures in at under 90 calories.

Immediately after ejaculation the male body is in no condition to do much. While the second set of autonomic nerves are firing away to

Orgasm switching center: abdominal ganglion of the sympathetic autonomic nervous system, through which the signals that herald the switch from arousal to orgasm must pass.

produce ejaculation, the first set, the ones that were behind erection, are dwindling away precipitously. Once the fireworks are done, this is likely to become only too clear. During the orgasm the brain apparently loads up with a chemical that's shaped almost exactly like morphine. That's probably what's behind the feeling of effortless well-being after orgasm, for this now plentiful natural chemical will fit into, and trigger, every one of the pleasure-producing and pain-canceling nerve centers that concentrated morphine would. That part of it is great. Unfortunately, the same chemical is also quick to swirl into just that part of the hypothalamus in the brain where the LHRH hormone that was behind sexual appetite gets produced, and block it cold.

Barely 33 percent of men under 25 report success in having multiple erections and ejaculations, while at higher ages this percentage drops precipitously. It's not just that the efforts to return to the original nerve channel of arousal proves more than aging bodies can take, or that the necessary reworking of the plumbing is hydraulically impossible either, the interest, due to these chemical changes in the brain, is just not there.

That's why suggestions can be as important as physiology here. While American men average about two orgasms a week at age forty-five, men of the Nungauan people in Polynesia, who have just the same physiology, reliably report having two orgasms a night, every night for several decades after puberty. The difference perhaps is that the Nungauan put much of their social value on the ability to please others with frequent and active sex, so that interest to counter the LHRH-canceling chemical is maintained, even when the novelty factor has worn off. The occasional solitary bursts reported by the Americans fits in nicely with a society where so many objects are designed and marketed to satisfy an individual user, with little consideration for the consequences on others. It's the ultimate predicament of the ultimate consumer, with everything going to serve him, and he unable to serve any other.

Women (continued)

We last saw the vagina and vicinity well into their arousal, with the inner lips or labia minor having just changed from a dull pink to a glowing red, or even on to a dark wine color. It is a sure-fire sign that an orgasm is soon to approach, and it precedes changes in the rest of the system that are markedly different from men's and that help explain the more diffuse, often more powerful orgasm that women report.

With continuing arousal the whole shape of the female tract changes. Organs lift and tissues reposition themselves as the neural ballet of whizzing signals deftly does its stuff. The uterus starts lifting up from its normal position bent over above the vagina and ends hoisted up by internal cords to a much less sharp angle with the vagina, reaching a level above the navel. Any general feeling of movement in the lower abdomen, then, even well above the vagina, is not in the imagination. The uterus's entry down into the vagina, the cervix, also goes through a major repositioning, and after some aimless flopping from side to side, draws back tightly as arousal continues, so drawing out the tubelike length of the vagina. Tubelike for long it does not stay.

The whole vagina is wrapped in strong muscles, that are sometimes called slow-twitch, for they are more ponderous than the ordinary arm or leg muscles, do not start and stop so easily, and are not under direct conscious control. Similar slow-twitch muscles are also found around the intestines, or controlling the lung's steady movement. Their sliding microfilaments are caught inside the cell's viscous protoplasm here: in the ordinary muscles they're arranged in the long bundles we saw earlier.

Yet despite their limitations, the slow-twitch muscles are strong. Along the upper two thirds of the vagina they automatically balloon out as arousal approaches its peak — loosening the vagina there to protect it from possible bruising. The lower third of the vagina, where most of the sense receptors are concentrated, gets the opposite treatment and is battened down snugly. It turns into a narrowed circular channel, guiding any incoming objects on a relatively straight path, and so further reducing the chance of unwanted impact on the walls of the upper vagina. The tightening also brings its own nerve endings into a pleasant contact with whatever might be being guided.

The place of breasts in all this deserves a mention. Breasts enlarge mightily during sex: up to 30 percent in many women, especially those who haven't given birth. Nipples swell up too, with growth of over a quarter-inch common, though here the tyranny of the cylindrical nipple must be firmly laid to rest. Nipples come in a variety of shapes: there are cylindrical ones, mushroom-shaped, cone-shaped, fissured, plateaued, wrinkled, flat and — not at all unheard of — pushed-inward ones. Their changes during excitement are likely to be equally idiosyncratic. Very wrinkly nipples are due to the undersurface of the skin on them being bunched up in spongy ridges, few ridges, fewer wrinkles. The only common ground is that almost all women's nipples have 15 or so tiny crevices in the tip (usually filled with horny plugs), and at the peak of arousal are so sensitive that they will not take kindly to overfrenzied pawing. For the breasts are inherently delicate structures, chock-full as they are of fat, nerves, lymph fluid, arteries, connective tissue, milk glands and ducts. They can often be intensely erotic, but only if handled properly. Over 35 percent of women in one detailed English study reported frequently masturbating to orgasm primarily by extensive breast-fondling alone. The key was that it was under their control, and modulated quite finely when sensory etiquette demanded. That all sexual partners should have such grace.

Just before orgasm it might seem that the vagina's fluid secretions have increased, but that's only an effect of the long buildup; they're actually continuing at just the same rate as before. If the ballooning on top and the clamping on the bottom are going on during intercourse, one or two drops of fluid from the penis are likely to have joined them before ejaculation. These drops are produced by the small Cowper's glands tucked below the base of the penis. Dribbling usually unnoticed into the vagina, they produce an extraordinary change on its internal chemistry, for the usual lubricating secretions there are

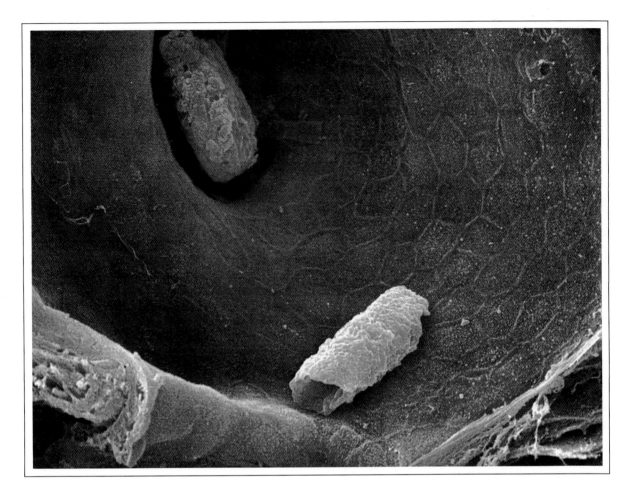

Moisture being released through vaginal surface in arousal.

slightly on the acid side, and though this acidity is too mild to hurt the skin, it does have the effect of killing most of the sperm that get within its touch. Those not desirous of having children with each sexual experience might think this all to the good, but the body has less discretion in the matter. These drops from the Cowper's gland mix with secretions in the vagina and counteract the acidity, undergoing the same fizzing reaction as when a glass of bicarbonate of soda is swallowed to soak up the acidity in a troubled stomach. The alkaline drops change the taste of the vagina's outpouring from a slightly salty one in the early arousal to a much more bland taste just before orgasm.

Other changes are going on, too. Just outside the entrance to the vagina, the usually sensitive clitoris is by now swollen by signals from the same pelvic nerves, acting with the same acetylcholine transmitter, that squeezed down on the veins around the penis to produce the erection. The clitoris has all sorts of empty little chambers inside, that when filled with blood will swell and swell. This is only to be expected, for both clitoris and penis (outer lips and scrotum too, for that matter) originated from the same bud of tissue in the fetus, going their separate ways only in the ninth embryonic week, just about when the first nails grew and the eyelids first delicately closed.

The swelling produces such extreme sensitivity on the clitoris that it slides its one inch length back under its hoodlike covering at the top junction of the inner lips, so it's barely visible. There it's protected from any too harsh grazing, but cunningly so, in a way that leads to even more arousal, for the hood over the tip of the clitoris gets tugged back and forth by any movement nearby, be it bodily impact on the hair-covered pubic bone just above, or slipping and sliding in the entrance to the vagina just below. Through the intermediary of the hood these far too large-scale actions are geared down to a precisely measured amount of stimulation, an amount that the clitoris itself changes by the simple expedient of slightly increasing or reducing its swelling to keep just the right distance from the strokes of the hood.

This fine calibration will be totally thrown out of kilter by any sexual partner who, believing that the clitoris is the center of female delight, proceeds to search it out with Pavlovian fervor while the poor tissue is trying to hide out of direct touch under its protective hood. Like ordinary skin, the now deep red clitoris has tens of thousands of nerve sensors, swollen into maximum action. Some are merely tuned for detecting warmth, cold or touch, but there are others that when stroked against signal unpleasantly that there is pain.

When not unduly harassed, an aroused body will see to it that its clitoris gets even more precisely regulated amounts of stimulation. That's because the clitoris is held firmly in place in the tissue between the inner lips by two stretched cords which pull like taut guy wires all the way down to their support on the pubic bone. (They're called the *crura* from the Latin for "leg," *crus*, because they're wedged tight like straightened legs. The two crura are themselves surrounded by even tinier muscles, which can be contracted to pull down on the clitoris *from underneath*, producing a fresh sensation. The curious conclusion is that the erection of the clitoris all women get during sex is almost never due to direct stroking by the penis, or any other available appendage. Far more useful is fantasy, breast stimulation, pressure on the mons veneris, or stimulation of the inner lips or rectum. A halt in proceedings this far along will be intensely frustrating, not least because the blood gushing into the clitoris to give it its erection also leaks some of its plasma components into all the surrounding genitalia and other tissues. That makes them swell too, just as in edema, and unless an orgasm is forthcoming the out-of-place plasma is going to keep them swollen and sensitive, until it all, and only gradually, leaks back to where it came.

Since the clitoris is small, and the muscles pulling it are smaller still, even at full contraction, squeezed tight, they will not pull enough from underneath for full delight. They're too weak. How to aid the tiny muscles tugging on the crura, which are tugging on the clitoris? Arching the back will do it fine. This move comes about because the brain is getting bursts of nerve signals from the clitoris saying that the little reverse pulling it's had is a great thing, and that more would be greatly appreciated. *Now.* The brain can do its damnedest with more signals from the spinal cord to the muscles around the crura, but if they're as tight as they can go there's little more help to be gathered.

That impasse leaves only the indirect approach that does the trick. Other signals go out from the brain and spinal cord, subliminal and quick, signals to apparently unrelated areas such as the back of the shoulders, lower back, and outer thigh. All called at once, they'll all contract at once, and that's what produces a real arch in the back. In that position all the pressing on the pubic bone, all tugging on outer lips, above all much of the sliding inside the vagina (which doesn't have many nerve endings itself), will get redirected with much more force on the furiously straining crura attachments underneath. That means more pulling from under the clitoris, and more of the sensation. Inside the now strainingly erect clitoris there are hundreds and hundreds of movement detectors that wrap around the branching blood vessels in there and can detect movement of under 1/1000 of a millimeter. That's why every extra bit of back-arching can be so delightful for the doer.

While arching like this is a reflexive change, coming without thought near the extreme point of sexual excitement, it uses only muscles which can be called on at any time. The shoulders, back and thighs needed to arch the back can be pulled taut by conscious intervention. Soon, though, come other changes which can't be brought on consciously; all led by the switch to the other system of nervous control that causes orgasm. The first system, remember, is the one that worked by spurting acetylcholine along to contracting muscles, usually the ones that control the veins out of a tissue, thus producing swelling in them. Sometimes the ear, nose, breasts, bottom of tongue and even toes will undergo some swelling when this system is working fully. In the genitals this system is controlled by the pelvic splanchnic nerve, which snakes out from the lower spine; elsewhere, especially in the face and breasts, its swelling actions are controlled by top-of-the-line cranial nerves, which lead out directly from the bottom of the brain, sometimes through special holes in the skull.

If this nervous system just went on with its signaling, the swellings associated with the arousal side of sex would go on indefinitely. Pleasant, perhaps, but after a while unnerving. The feeling that it's too nice and one must stop is likely to come about whenever this happens for too long. Only with the switchover to the second network of the autonomic nervous system does the expected sensation of orgasm occur. And that switch is not instantaneous. A study of seven young women who were masturbating till they reached orgasm showed that the total volume of blood in the body has to take a sudden drop before the final stage can begin. The brief moment of transition is what produces the frequent feeling of anticipation. It ends, of course, when the second system is in command, for that heralds the orgasm itself.

The uterus tightens deeply, a change pleasant in itself, which also brings closer to the fallopian tubes and any waiting egg, any sperm there might be. The muscles around the vagina tighten in three to fifteen contractions spaced .8 seconds apart, often centering the delightful compression at that point a third of the way up the vagina that was the limit of the tightened section before. The muscles feeding into the clitoris (and also those around the supporting crura) pump back and

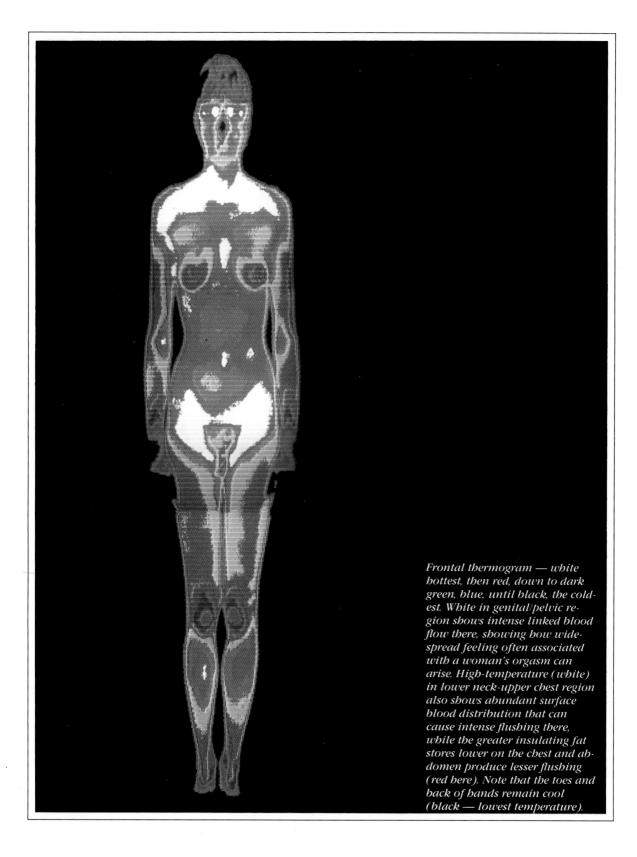

Frontal thermogram — white hottest, then red, down to dark green, blue, until black, the coldest. White in genital/pelvic region shows intense linked blood flow there, showing how widespread feeling often associated with a woman's orgasm can arise. High-temperature (white) in lower neck-upper chest region also shows abundant surface blood distribution that can cause intense flushing there, while the greater insulating fat stores lower on the chest and abdomen produce lesser flushing (red here). Note that the toes and back of hands remain cool (black — lowest temperature).

forth in spasms too. And the bulbs of the vestibule, those sensitive tissues just below the clitoris that curved around the vagina opening, also are surrounded by tiny muscles, and also get deliciously scrunched tight by the fresh nerves. Further squeezers, exactly in time with the orgasm's pulsing, are the anus and the lower abdomen.

Once it finishes, there's no need to build up the process laboriously before another surge of firing can start the orgasm again. Three to five repeats in a few minutes is often reported, and ten or more in an hour have also been noted. Although men have to go through the often-long process of reerection with the first nerve system before getting an effect again from the second system, women do not. And for some reason the possible problem men have with the natural morphine outpouring after orgasm that blocks libido, seems not to arise here. For a woman all the delights of orgasm happen on the bottom of a sheet of muscle that indirectly wends its way up the center of the abdomen to the level of the lungs. The outer labial lips, for example, are hooked on underneath to a sinuous ligament that stretches inward all the way to form a ring around the uterus bottom, then on beyond to the very back of the pelvic bones. The swelling and tightening at the bottom has effects all the way up, and that makes for the broad, diffuse spread of warmth with which the whole process is often so fondly recalled.

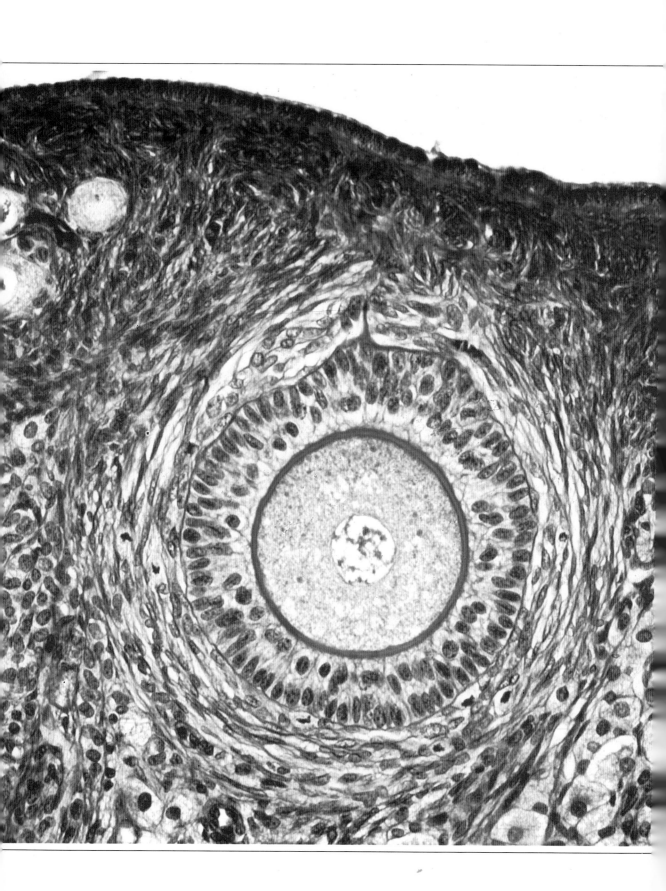

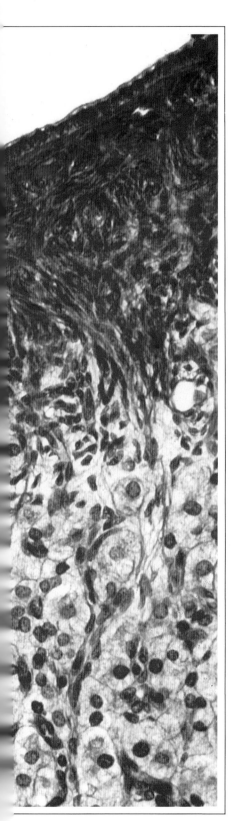

4
Conception and Pregnancy

*U*sually, after sex, that's it. A bit more sweating, comments sotto voce or otherwise, a lengthy or hasty disengagement, and the two partners are back where they started. But usually is not — as the $130-million-a-year pill industry, multimillion-dollar IUD industry, and multibillion-dollar-a-year baby-clothes industry all can testify — quite the same thing as always. In an American average of once every 1,500 or so times, the woman becomes, as the ancient Greeks so poetically put it, "soon to germinate," a phrase that came out in their language as *praegnans,* and which is mispronounced in ours as "pregnant."

Fertilization

The standard picture is well known. The female egg rests, eager but passive, in the fallopian tube, while the male sperm swims furiously all the way up to it, and one especially forceful one batters its way in. It's the picture most of us learned at school but it is, alas, totally false. The egg does not wait passively, and the sperm does not swim all the way up to it. Nor does it batter its way in once there. The standard description, in fact, was developed in the mid-1800s, by bourgeois professors in Germany, and the pattern they described bears a suspicious resemblance to the proper course of middle-class marriage so respected at the time. This notion of female passivity and male domination, as acted out even on the level of sperm and egg, was shown to be false as long ago as the late 1940s, when the first electron microscopes were applied to embryology, but the new discoveries have been slow to filter down from scientific papers into classroom textbooks. This is a lack we shall remedy here. Dispose of the image of fertilization as a conventional marriage; instead think of it in terms of a 1950s high school prom.

First there's the egg. It's a remarkably homely object. It would never leave the ovary of its own free will, and in fact can be seen through a microscope hiding, all squinched down, in the very center of the ovary until its hand is forced. Pushed up to the surface exit by a swelling bubble, and gets flung out of the ovary through no choice of its own. Now it's in a dilemma. Stretching before it is the menacing channel of the fallopian tube, which leads on to the uterus, where the sperm might be entering and where all the action is. So it just waits there, halfway between ovary and fallopian tube, stock-still, and not daring to move.

Things would stay like that but for the uncouth vigor of some stunted hairy cells sticking out from the entrance of the fallopian tube above the ovary. They wave. At first the egg does not move and in fact it can be seen in the microscope, standing stock-still as the stunted sentinels gesture. But soon, the wavers at the front of the fallopian tube start straining so furiously that strong currents are set up in the fluid around them, currents that reach where the egg is waiting. They swirl around it, swoop it up to their station by the gates of the fallopian tube, and without a pause send it all the way in.

Even with this forced entrance the egg falters into one of the drape-like folding undulations that are hanging all around the edges of the fallopian tube. It hides so well in fact that for three centuries after Fallopius first discovered these channels from ovary to uterus in the mid-1500s, no one was able to pick out the egg hiding in the folds of the wall. Poor Gabriello Fallopio had no idea of what existed in the tubes he had discovered. But why, one may ask, is the egg so hesitant about standing out in the fallopian tubes, or going on to entering the scene of all the action further down along in the uterus? Couldn't it just blend in with all the other cells of the body, and if not exactly boggie, then at least do a shambling imitation of the latest moves? Alas, it has good reasons for being so shy. For the recalcitrant egg is, to tell the truth, perfectly ugly. First of all it's fat. Not plump, not a little bit chubby, but gapingly, gaggingly, enormously, fat. By this time, the egg cell is, after having been healthily nourished back in its ovary home, a whopping eleven times heavier than the next largest cell in the body.

Imagine the school's fattest girl who was forced out of her house to go to the dance, only came in because she was pulled by uncouth kids by the entrance, and then realizes that she has, horror of horrors, left on all the cold cream and the hair curlers she put on at home. That's exactly what our ungainly egg cell has suffered. In fact, the egg cells seems almost stuck, for protruding out from it, all over, are lots and lots of tiny, misshapen spiral protrusions, called epithelial cells, that clung to it when it was leaving the ovary. And under that is a thick, creamlike layer that by now is warm jelly. The shame of it. It's certainly enough to make a wallflower out of anyone. That's just what the egg becomes, tucking up as best it can behind a curtainlike fold away from the dance floor, hoping to stick out the ordeal until finally, hopefully unnoticed, it can get to leave.

Fully formed egg being propelled to surface of ovary for release. (overleaf)

Ovary in a three-year-old girl (above), and a twenty-seven-year-old woman (below) showing pitting due to egg release each month. The original ovary contains several hundred thousand proto-eggs; fewer than four hundred of them will mature and be released in a lifetime. The fallopian tube is above.

There for a moment we shall leave the egg, and switch our attention to the other end of the genital tract, the vagina, where, entering the high school from the other direction, as it were, are all the favored sperm. These arrivals are just like eager juniors getting their first look at the senior prom. They're ungainly, having only made it to their full height a few days before, and so shy that they don't enter one by one, but are only at ease by blustering in as a group. And in they swarm, not two buddies, not a dozen club members, but an anonymity-ensuring bunch that can be 400 million strong. In that rush, swaggering tall in the middle of them all, moving in step with the others but still twitching nervously to each side, is the one who will be our subject: the most Typical Sperm, T. Sperm. Although it doesn't know it yet, T. Sperm is the one that's going to reach the hiding egg. Considering that it's three times shorter than the egg, and over thirty times thinner, it's perhaps best that it doesn't yet know what's in store.

Like all the other sperm, T. Sperm moves by the peculiar expedient of keeping its lower body stretched out long behind it, lassoing a ropelike cord tight around it, and then tugging on the front part of the rope with its torso. It's like stepping into a burlap bag for a sack hop, then hopping forward by tugging on it with a rope. This twists its lower body enough to skid forward. The movement isn't fast, well under ten inches an hour, but considering that all the other sperm lumber around in the same way, that's just about right to keep in place. Now aside from twisting tail end, lassoing rope, and torso, there's not much more to T. Sperm. Up on top he is frankly expressionless, like all studiously nonchalant adolescents, although graced with a whopping great amount of DNA filling up his head, storing enough information to write dozens of *Encyclopedia Britannicas*. This is the setting — a behemoth hiding egg, a ridiculous, barely mobile sperm — that is supposed to fulfill the standard picture of male capture of the female that the biology textbooks still suggest.

Anyone with a firsthand knowledge of school proms could point out the problems with it. Is it really likely that the grossly overweight egg, still with its curlers and cold cream on, would come out from its wallflower hiding? What's more, is there any chance that T. Sperm, desperately trying to be just one of the boys, would suddenly leave all his buddies behind, fall in love with the hideous apparition, and then have the wherewithal to consummate his passion? Doubtful, truly doubtful. What really happens at the start of pregnancy is much truer to a script written on sound feminist lines: it turns out that the male sperm is the real passive object, while the female body and egg are the ones that take control. One single male sperm does not suddenly get the urge to seek out a mate. Instead the whole uterus pulsates inward, in great pulling grasps, during the moments of female orgasm. The cervix, leading into the uterus, cranks wide, stretches down into the part of the vagina where most of the sperm ended up, and whooshes them up in vacuum-cleaner-like slurps.

That's how T. Sperm ends up in the fallopian tube — he doesn't have to leave his compatriots, but instead gets whooshed into the tube along with vast numbers of still-bobbing, expressionless chums.

Sperm in a frenzy, fleeing every which way they can. Note the tangled tails and random direction of movement. If the number of sperm in an average ejaculation were enlarged this much they could form a chain around the Equator.

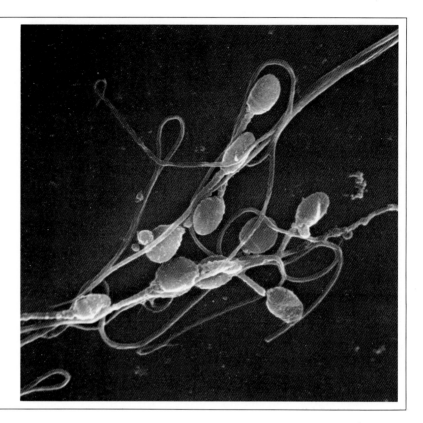

(Sometimes, of course, the sperm can swim the whole way without this help; that tends to be slower.)

After all, they've been mysteriously swooped by a suddenly opened coal-shoot into an unsuspected sub-basement. Only this time the coal-shoot leads up, instead of down. What the sperm do first in the fallopian tube is predictable. They scatter and rush away blindly in all directions. This rushing is not efficient, however, for they're all still moving only by tugging on the cords leading down to their lower bodies, and wriggling around as well as that lets them.

In this blind free-for-all, this frantic running, one of the sperm gets the shock of its life. Wishing only to desperately flee, it smashes smack into the curler-covered, shamefully hiding egg. This is no love-sought meeting, but a randomly produced, blind collision. Nor does this first sperm act like it's in love, for upon contact the sperm's first action is to try frantically to wriggle away. In it's distress it also sends off a chemical call to all the other blindly moving sperm in the vicinity. Those that pick it up head on over. One by one they too crash into the egg. One by one they too start to wriggle as if trying to escape the egg's clutches. Among them now is T. Sperm. The flailings don't do much good; instead they serve to get the dozens of head-caught sperm stuck deeper in the spirals and creamlike jelly.

At this point, with the awful shrimpy juniors pressing hard against the wallflower of an egg, something very beautiful happens. In this indelicate situation the now totally humiliated egg suddenly discovers, dare one use the romantic phrase, the hidden depths of a woman. The ugly duckling is no longer. Suddenly, and very majestically, the egg brings out from her depths an innate charm (admittedly a chemical charm), that makes the sperm now less desperate to just struggle to get away. It distills out a wonderfully transfixing spray. Suddenly the sperm struggle to see what's inside. Is it an angelic goddess who resides under that earthly garment of hair-curlers and fat? A wondrous princess, cruelly misshapen by an unkind and maligant nature? Whatever it may be, they go at the job with a new aroused desire.

Perhaps too much. In their freshly-formed excitement, and revealing their barely adolescent nature, the sperm do a most embarrassing thing. Perhaps on the principle that little fleas have littler ones, the now aroused sperm around the egg suffer the shame of a group premature ejaculation. They all start squirting out from their foremost parts a thin, sexually useful, almost milky fluid. Shame. Only the egg, in her new-found maturity, is not embarrassed by this sign of overheated ardor. For the fluid is going to help bring the egg cell's control of these suddenly sex-craved sperm to a close. Although milky, the fluid is also slightly acidic and is enough, when pulsed out in a group endeavor by the dozens of affixed sperm, to wear a path gently down through the curlers and cold cream, down through the most corpulescent surface, right into the egg's now ever so desired center.

The opening path is deepest first on one side, then on another, but only under T. Sperm does it finally go all the way. The fluid-bored hole creates a tunnel into the now lovely object of desire into which T. Sperm swoops down, with such excitement that he even leaves his torso and body quivering emptily behind. That's a headlong flight if there ever was one. Once he's in, slam, the egg, with her newly found savvy, reflexively knows that the only way to keep her new suitor is to close the hatch behind him. The jellylike cold cream layer suddenly thickens and solidifies. Just to make extra sure the egg also sends up tiny boulders, which it had lying like pimples all over its body just under the surface, pumping them up to the relative size of beach balls, and popping out these now rock-hard defenders all over its outer surface. That definitely is enough to keep the now-bodiless T. Sperm in, and as a secondary result, it's also enough to keep the other still enamored sperm out.

Unequipped to ponder on the fickleness of desire, they wriggle some more, only to soon give up the task in exhaustion, and die one by one, there in the isolated depths of the fallopian tube. They will never get a chance to take a second look at the path they traversed. T. Sperm and the now securely encompassing egg however do. Locked together in their safe covering the two linked souls float back over the path the sperm horde was sucked along through the fallopian tube. The couple pass uncaringly over the strewn and mutilated bodies of the other pulled-in sperm, a scene of utter death and devastation. It's a cold and uncharitable view this floating pair take, but it's one that does perhaps satisfy the feelings of total revenge that can be so readily appreciated by all who have ever been unloved wallflowers, or nervously strutting junior jocks out on the make at their very first prom.

Genetics

This meeting of sperm and egg happens unnoticed to all the outside world. The maximum probability is perhaps 19 hours after intercourse (extremes of 30 minutes, and 2 days), which, assuming a late-night endeavor, works out to something like 6 P.M. the next day. What happens next, another 12 hours later, around 6 A.M. by the odds, goes just as unnoticed. The genetic inheritance carried in the two cells mixes together to create the blueprint for a new person. How exactly it happens is a subject about which there is much unnecessary misunderstanding. One problem is the unexplained use of statistics in this enterprise. Whenever the newspapers announce that some unsuspecting woman has given birth to quintuplets, for example, they are always quick to point out that this is somethng which has a probability of around one in 20 million. It leads one to wonder how the poor woman had time for anything else.

Early thinkers found the problem no less complex. For a long period in the 1600s advanced scholars at Oxford and the University of Paris were convinced that a baby was conceived because the sperm cell had inside it a fully formed little person, quaintly named a *homunculus* (Latin for "little man"), which popped out of its transporting module once it was in the womb, and then grew and grew until it was time for it to be born. The problem was that there would have to be an infinite series of littler and littler persons all packed inside each other, to provide the source for yet future generations. Researchers were stuck with the notion that children were somehow a 50-50 mixture of their parents.

This made sense, perhaps, at a time when feuding kingdoms were still being united by judiciously arranged royal marriages, with the union of the two royal houses designed to symbolize the union of two lands. It was considered normal, for example, for a fourteen-year-old Austrian princess to be sent off to Spain to grace the bed of a lecherous fully grown Spanish king, all for the purpose of blending their respective subjects. Only after such forced marriages died down did it become easier to look at how much a child is *different* from its parents. That was the real clue. It was certainly helped by the writings, in the late 1700s of the early Romantics, who extolled the virtues of childhood and made a big point of showing how much a child was an independent person, who should be treated with the respect every person deserved. Their formal views backed the idea that children were *not* merely low-grade blendings of their parents. It's an idea that has common sense behind it, for if genetics did work by making children an even blend of their parents, we would all gradually come to be the statistically average American.

How then does a newborn infant come to somewhat resemble its parents but not be exactly like either of them? The first stab at the modern answer came with the researches of an Austrian monk named Gregor Mendel. Although now enshrined in all the histories of science, he was known at the monastery where he lived, in the mid-1800s, as just a portly fellow with an odd interest in vegetables. What

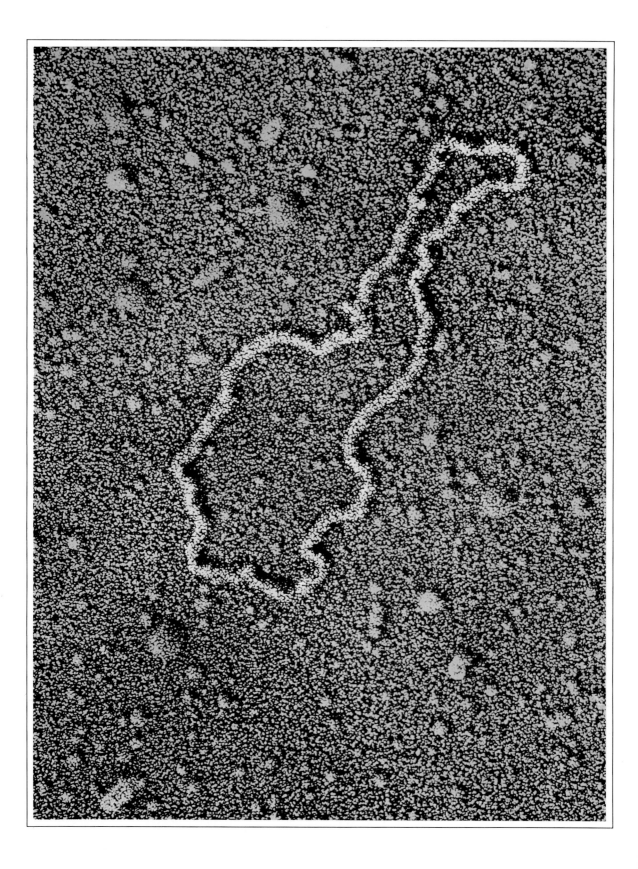

DNA. That strand contains more information bits than a good-sized home computer. If an ordinary necklace were enlarged as much as this micrograph it would fit around Manhattan.

Mendel did was to carry out experiments with peas in which he saw that the offspring of two different plants would sometimes inherit all the characteristics of one parent and sometimes all the characteristics of the other. What was more, the offspring fitted into one of these slots according to what the ever-stolid Mendel thought was a series of precise mathematical laws.

Buoyed by the discovery he sent a copy of his results to a leading geneticist of the time, who unfortunately was not much past the stage of believing in homunculi, and who never deigned to answer the letter from the obscure monastic tenant.

Poor Mendel was back with his peas with no scientific succor insight. As a final report he published his findings in an obscure local journal, where they remained, untouched and unread until a full fifty years had passed. He also published figures that are a lot more precise than modern knowledge shows they had any chance of being, though the conclusion that he fudged them to gain attention must be tempered by the knowledge that he was a monk, and so perhaps had special assistance.

When his findings were later picked up, it was clear that Mendel had been smack on the right track. To produce offspring that are different from the parents, there must exist cell-inhabiting little chemicals of inheritance, one from each parent, which sometimes mix to produce one parent's characteristics and sometimes that of the other. This chemical controller is the much famed DNA. In each of the body's 100,000 billion or so cells, there is a set of pairs of helical strands of DNA, called chromosomes. As we shall shortly see, one member of each pair was inherited from the person's father, and the other from the mother.

Coded in the DNA are the complete instructions for how every cell in the body should be built. Most cells have no use in their ordinary tasks, for the complete listing, and so slip a special barrier over almost all the DNA they contain that covers up all the DNA except for the brief bit controlling how that particular cell is to be built. For example, in nose cells there is a covering over all but the little DNA bit that has the rules for how nose cells can be built, while in finger cells, the nose-outlining bit is covered, and the only section of the DNA strand that's left clear is the one saying how finger cells are to be kept in shape.

Only in one set of cells in the body is there only one set of chromosomes, and their DNA is easily uncovered. These are the sex cells: sperm in men, eggs in women. Once the two have joined, or, more accurately, once the sperm has been pulled into the aggressive egg, the DNA strands they each contain are in a position to divide up the inheritance turf. Before this can happen, one might think there is a size imbalance to take care of. The sperm, remember, was some 30 times smaller than the egg. But the DNA it carries is no smaller than the egg's supply, for with such a disparity there would be no way they could join. Still, neither is enormous: the DNA of all 4 billion people alive today could fit easily on a teaspoon and would weigh less than one twenty-fifth of an ounce.

Approached together, the sperm and egg DNA set the blueprint that will lock in so much of the resulting offspring's destiny. While it's going on, of course, the woman has no idea that this egg out of all the others has been fertilized, nor does she know that the DNA heritage of the offspring is about to be set. For her the first inkling will only come several weeks later, when the first period is missed, and the breast tenderness, weight change, and other signs of early pregnancy set in. That's long after the male sperm's DNA within the egg approached the translucent chamber at the very center of the egg where the woman's uncovered DNA store is resting. There the two genetic strands twist into mirror-image patterns.

This lining up happens with the two strands close, but not touching. Soon that changes. Perhaps while the woman is picking up a phone receiver, perhaps while she's sitting back in a car seat, the slight electric charge on the two facing DNA strands stretches out over the slight distance between them, and pulls the two storehouses together. They match up point-by-point at each item in their body-outlining inventory. The part of the sperm DNA that in the man's body went into controlling the nose cells, here fits up neatly against the part of the egg's DNA that in the woman's body would also be found active in the nose cells. All the other items match up in just as precise a fit, knee-cell instructions against knee-cell instructions, eye-color ones smacking in touch with eye-color ones.

This combining presents a serious problem. The resulting child would get no benefit from having a double set of nose instructions, as well as a mix of eye-color instructions and the rest. Somehow the two so suddenly linked DNA molecules have to divvy up the job, deciding which one's instruction will be followed for each cell in the child's body that's to come. The way they do the sharing is comparable to the way in which the children's game of paper-scissor-stone is scored. In that game, when one player puts out the stone symbol, and the other puts his fingers out in the paper symbol, the one who took the paper's part wins, on the grounds that paper wraps around stone. Similarly, if one puts out scissors, and the other one stone, the one doing stone wins, on the ground that scissors blunt on stone. There's a preset decision of who will dominate for each one of the possible matchings.

That's how it works in the fertilized egg cell. If the DNA segment that produced brown hair in one parent matches up with the DNA segment that produced red hair in the other, the rule that both follow is that the brown-hair-controlling segment will win. That's the one that will be effective in running the development of the soon-to-grow embryo; the other one just sits it out. For all the other possible linkings, snubnose against straight, green eyes against blue, there are preset rules on which one gets to predominate in the embryo-to-come. Clear winners are long eyelashes over short ones, curly hair over straight, and early baldness over continued sprouting. There are, of course, times when it's less clear than that. The shape of your eye, for example, was determined not by one, but by a whole bunch of DNA slots for matching up. There were some that fixed the amount of earlobe sagging, others deciding the width of the ear, the shape of the folds, and how much it sticks out etc. But in each case the divvying-up principle and the preset rules to avoid dickering between the parents' DNA is the same.

This, incidentally, provides an answer to the old question of which came first, the chicken or the egg. It was the egg. Once its DNA was set, there was no question but that a properly gabbling chicken was going to come out. The mother in this case was not quite a chicken, possibly an aardvark with wings, more likely something like a pigeon that laid a lot of eggs. The DNA for the egg got changed from the mother's pure form, probably by a peculiar divvying up with the male mate's DNA, possibly by the odd cosmic ray mutating it on a bit further, and the result was the world's first chicken.

Usually both human parents supply characteristics, but in one notable exception, the input of just one parent determines all the field. This is the matter of what sex the child will be. At the point on the mother's DNA where she has the instructions of what sex organs and hormones to make, only one type of DNA instruction is marked in: feminine. The mother can code only for daughters. Not so the father. Matching up to this space of the sperm's DNA there are two different kinds of sex instructions he can have provided. Some sperm have the instructions to make a boy there, while others have the instructions to make a girl. If the sperm with the female description was the one that got into the egg, the result is going to be female; if it was a sperm with the male description, the result is going to be a boy.

This means that in any cases of virgin birth that might arise, where for some reason the female egg starts to double its DNA and multiply on its own account, there is no question but that the offspring would have to be a daughter. For when the egg's DNA doubled up to produce the first cell division at the start of the whole proceeding, it would automatically supply a DNA segment coded for femaleness in the slot for "sex of child." That's the only DNA code it has on the spot, since only DNA from a male sperm could fill in the masculine-producing DNA segment needed to produce a boy. This makes historical accounts of virgin birth of a daughter believable, if unlikely, while rendering any accounts there might be of virgin birth of a male less believable by far.

If the English King Henry VIII had been trained in modern molecular genetics like this, he might have restrained his notorious temper when five of his six wives did not produce a male heir for him, and instead have said something like: "darn (or darne), my insufficient male-coded spermatic DNA seems to guarantee me only daughters." (But he was not of the type to be satisfied with such recondite observations, hence the intemperate divorce of Catherine of Aragon and Anne of Cleves, then even less temperate beheadings of Anne Boleyn and Catherine Howard, and perhaps in just recompense, the utter lack of lamentation of his sole surviving widow, Catherine Parr.)

It's in areas like this — sex, eye color, and hair color — that genetic determinism is as strong as people believe. Some of the things the researchers have worked out are remarkably precise. For example, all the children of a marriage between two color-blind parents will be color-blind, while if only the wife is color-blind, the sons will be color-blind but the daughters won't. That's exact. Unfortunately, few of the interesting, complex traits a person has are simply determined at the moment the two DNAs first sidle together in the egg in the fallo-pian tube. Instead they're the result of long and much less clear-cut matching of this home environment against that, this food at a tender age vs. another.

Some people say that IQ is an exception to this, and that it's an in-tellectual capacity that is determined with the sperm-egg DNA link-up itself. There is no evidence for this, but that was never enough to stop its most influential proponent, the late Sir Cyril Burt, long-time profes-sor at University College, London. Sir Cyril's deepest belief was straightforward: the English ruling class is the most intelligent class in the world. With this as a conclusion, the rest of his work followed eas-ily. In an impressive range of books and learned articles he proved that Jews were less intelligent than Englishmen, that Irish people were less intelligent than Englishmen, that women were less intelligent than Englishmen, and that Negroes, well, really. Within the noble order of English manhood, Sir Cyril was able to detect a significant distinction: the unskilled classes were less intelligent than the higher professional classes. Unpalatable though this might be to those of tender con-science, he wrote that it was nothing but clear, unambiguous fact.

Sir Cyril's greatest popularity came in the 1940s and 1950s. He based his findings on what he said were a great number of intelli-gence tests carried out on different people. He never showed anyone those studies, but his conclusion that intelligence was innate and un-changeable was taken as a scientifically proven result — entering America in part as the abstract-reasoning portions of the SAT and GRE exams. Even when developing research on DNA put in doubt the con-clusion that intelligence was so fixed at conception, there was always Sir Cyril's massive evidence to conjure with. The bubble didn't burst until a few years after his death in 1971, when a few curious American professors looked into his data on file at University College and found that Sir Cyril's numbers had been largely picked out of the blue. The great man had been a charlatan, and the tests based on his work were unfounded.

Everybody has heard of the studies on identical twins separated at birth. Many of them come from Burt. He described the subjects of these studies as having been rent apart at birth and sent into obscure and widely separated portions of the globe, only to evidence the most extraordinary resemblances when reunited by chance years later. In fact, they were more likely to be two brothers in Manchester, one of whom was sent to live with an aunt a few houses down the street be-cause their own house was too small to hold twins, and both of whom grew up together, sharing their friends, clothes, schools and ideas. There is no inherited intelligence or personality that will show true in drastically different lives. All Sir Cyril's conclusions were lost. Blacks were not dumber, nor were Jews, Irish, or women and the tests that said the opposite had been rigged by Burt and his colleagues to that end. (But the Educational Testing Authority officials who make the SAT and other tests were too far entrenched in their products to go back to the beginning, and end them just because they have been shown to be based on a hoax.)

Burt's legacy, the whole notion of genetics determining all, lingers on in relations between races too. After all, black people look so obviously different from white that it seems only natural there should be deeper inbuilt differences too. It seems natural, but careful counting of the DNA shows it up. Every one of us has about 200,000 useful control gene slots in our DNA. It's easy to calculate from skin rejection rates between blacks and whites how many of them control skin color differences: 4. That's it. Just four slots out of 200,000 make your skin come out black or white, yellow or red. Most of the remaining 199,996 are spread randomly among the races. In this remainder are the DNA slots controlling the shape of your heelbones, the width of the main artery to your liver, the arrangement of nerve cells in your brain, and the like. There's a good chance you have more genes in common with someone on the other side of the planet than with someone down your street. There's a good chance George Wallace has more genes in common with Jesse Jackson than he does with Ronald Reagan. From a reluctance to do the counting most of our racial apologists draw store.

Implantation

The chronology so far: the egg captured the sperm 19 hours after intercourse (standard range 6 to 36 hours), and the two partners' DNA paired up, and set down the broad outline for the offspring-to-be 12 hours later. At that point the egg is fertilized. The name comes from the Latin *fertilis* — meaning "fruitful" — and is, poetically speaking, just what is about to happen.

First, though, there is a trip. For three days the new pair steadily drift along the fallopian tube, heading ever inward toward the uterus. They are sucked there by slowly pulsing muscles in the fallopian tube walls that automatically pull the egg along. The pair pass over the dead and dying husks of the millions of sperm that were sucked into the tube in the rush that brought in the one victor. These others were locked out of the egg once that sperm dived in, and they gradually died as their food stores ran out in what for them proved to be a fruitless journey. For the slowly developing sperm-egg couple, however, it is not that desolate at all.

In the fallopian tube is an all-penetrating solution of sugar water. It may have been useless to the hordes of frantically lost sperm but is now the ideal nutrient for the joined cells. Those cells soak in the nutritious juice through tiny pores on their surface and that helps them grow. This growth is not just in size, for the egg barely doubles its diameter in the three days' journey down the two-inch tube. Rather the growth is an internal one. The fertilized voyager builds up one new internal cell chamber after another, reaching a thoroughly labyrinthine two hundred internal rooms during the three-day glide.

Not all creatures spend their first three days of fertilized existence leisurely dividing this way. The noble pig, for example, has a fallopian tube 46 times longer than the human one, but it still manages to hurtle its fertilized egg down that channel to the uterus in a swift two hours. If the result is genetically set to be a pink, hairless, creature of porcine countenance, that's fine. The human egg begins to show its more refined different powers the moment it emerges from the tube. There at the end of its journey is the comparatively gaping cavern of the uterus. The fertilized egg at this point is just 2 mm across, smaller than the ball on a ball-point pen, and the uterus it enters is over a thousand times more vast, stretching in a roughly 2.5-inch-long hollow ball just below the navel.

Within this new territory the fertilized egg acts like an errant spaceship. Still without the mother having any inkling of what's going on, and now over four days since the fruitful intercourse, it careers about inside the uterus, tilting and yawning around and about in the apparently enormous inner space. It seems to be sampling the internal setting, like an overexcited scout ship speeding around in a new solar system. Sugar and salt concentrations are measured, and even thermographic readings of the uterine wall get taken by the swooping eggs all to see if there is a suitable place to land.

Sometimes its random glides take it too far off course — the minuscule fertilized spaceship swoops down too low and before it realizes what is happening and come back up again, slips out through the cervix, then the vagina, and falls, still unnoticed, clear away from the body. For some reason this particular mishap seems to happen more often to potential male embryos than to females. About 123 males are conceived for each 100 that are born, the rest slipping out early or otherwise spoiling the works before things get too far along.

Fertilized eggs that have a female genetic instruction locked firmly in place in their DNA, are much less likely to pop out or otherwise mishandle themselves. Forty-five percent of births are girls, showing the relatively sturdiness of that sex in the womb.

Males continue to demonstrate the feebleness of being the "weaker" sex once out of the womb, too. They are 51.5 percent of births, but 40 years later heart disease, accidents and other foibles will have switched the percentages around until there are more women than men of that age. The males steadily disappear more quickly throughout the years. Fifty-two percent of Americans aged fifty are fe-

What the inside of the uterus looks like when the fertilized egg attaches. Glands under the surface send up to it — under influence of progesterone from the cavity in the ovary from where the egg was expelled — glycogen, mucus, dissolved fat and other nourishing substances. Most of these build a fresh top layer ⅕ inch thick, which will be expelled, along with an average of ⅒ pint of blood, in menstruation if there is no implantation.

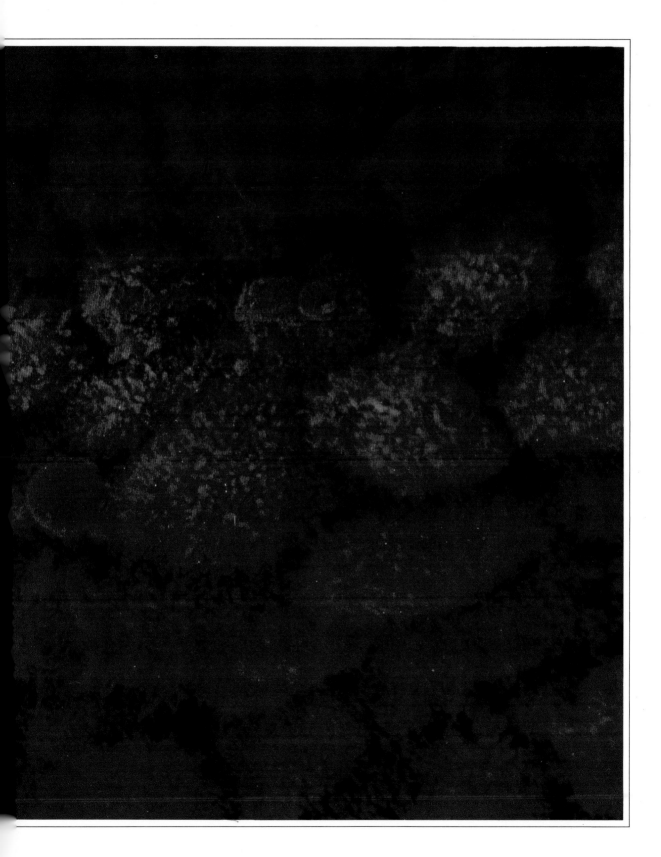

male, and 54 percent of those aged seventy. It only goes back to equal-
ity for the over-eighties; by then there are too few left of either sex to
keep much of statistical imbalance going.

But most fertilized eggs, after swooping around aimlessly in the
uterus for several hours, finally do slow down in their spirals and aim,
back-on, for a part of the wall about three-quarters of the way up towards the top, and ease straight in for an only slightly bumpy landing.
Three and a half days after conception and it's landed. This is where
the analogy with an explorer spaceship begins to break down. If anything the landing fertilized cell acts like one of the mutated spaceships
that figured so large in 1950s B science fiction films, the ones where
the American flagship coming back from Mars has been infected with a
horrible virus (symbolically communism? fluoridation?) that changes it
into a living organism that turns on innocent civilians where it lands.

The freshly landed egg in the uterus is itself half like a spaceship,
and half like a hungry parasite. The top side, the part away from the
landing ground, is a shiny half-sphere, glowing with a smooth gelatin
surface. But the bottom side, the one that touched down and is now
gingerly resting on the uterus's surface, is nowhere near as insensate
and sleek. Instead of politely waiting there on the uterus it immediately starts attacking and clawing out the living cells that make up the
top of the uterine wall just under it. The bottom of the implanted egg
spreads out in its unsuspecting host (The Blob That Ate Los Angeles?),
and starts chomping on the healthy cells it finds there, digging a way
to all the tiny blood vessels or protein storage chambers it can find.
Half an hour after it has attached to the wall, the top part of the egg is
still shiny and inert, but the bottom has spread out dozens of snaking
fibers deep into the nourishing surface of the uterine wall. From the
2 mm wide lander the digging-down supports reach five or ten times
that distance, out in all directions and deep into the wall.

That, incidentally, explains why men could theoretically get pregnant. The fertilized egg will start growing on any warm moist tissue
with a suitable blood supply. There have been cases of a fertilized egg
slipping out of the back end of a woman's fallopian tube, floating
around in the abdominal cavity until it latched onto the pelvic wall,
and growing there comfortably until it was taken out as a healthy baby
by cesarean section nine months later. There's no good reason, accordingly, why a properly implanted and hormone-supplied egg could
not enjoy the same normal nine-month growth inside a man. (Would
there be a problem finding volunteers to be the father [mather?] for
this historic event? So far though, the egg has only spread its feeding
roots in women.)

The probing growth is likely to provide the first notice to the
woman that she is pregnant. It makes a big enough change for even
the ovaries, by now four inches away, to start telling what's going on. If
the implantation continues, the ovary will send out diffusing hormone
chemicals to keep the uterus from being irritated by this furiously digging guest, that is also greatly stretching the uterus where it has
touched. If the implantation doesn't take place, or doesn't last after it
has started, the natural progesterone supply collapses, which causes

the whole inner lining of the uterus, rich in blood vessels and food stores just for the implanted egg contingency, to pack up and head out of the uterus in orderly fashion. That's menstruation (from the Latin *menstruus,* "monthly"), and its appearance accordingly is a sign that no pregnancy has been established this month. There are misreadings. The heavily settling egg can kick up a good-sized swath on the uterine lining, and the furiously burrowing cells on its base can send out even more damaged fragments of uterine lining. These bits knocked loose by the landing egg can give a semblance of menstruation, especially to someone who usually has very light periods, but in this case the semblance would be most definitely misleading.

Although informally called the curse, menstruation in previous times was sometimes called "the blessing," for it meant another guaranteed month of nonpregnancy, a state likely to be rare and greatly appreciated in some rural societies, where ten or more children might be the norm. One Russian woman has been reliably reported as having borne 69 children, a record composed of 16 sets of twins, 7 lots of triplets, and 4 sets of quadruplets. With that motivation, the topic of our next section should come as no surprise.

Contraception

The first contraceptives were designed by men to be used on women: it shows North African tribesmen tried to keep their women from getting pregnant by making them drink the collected froth from a camel's mouth. A venerable Sanskrit text insists that to avoid pregnancy a woman must drink flowers mixed with cow urine. And Baghdadi specialists of the Dark Ages suggested drinking insects with vinegar, while one European church authority held out for the woman's spitting three times in the mouth of a frog, eating the ashes of a wolf's penis and eyelashes, or both.

None of this suggests that the object of these ministrations was held in particularly high regard. In the one medieval case where birth control measures were decided by women themselves, ostensibly cloistered but energetic young nuns, such less unpleasant methods as magical blouses or special wristlets were chosen. (Less unpleasant but equally ineffective, illegitimate births in the nunneries were repeatedly condemned by church authorities, starting in at least the year 798.)

Men stayed clear of the trials involved, and even the condom was first used only when syphilis became something of an epidemic after Columbus's crew returned with this gift from the New World.

The same Fallopio who discovered the tubes bearing his name was one of the first exponents of protective sheaths, pointing out that they could easily be carried in a man's pocket for use on sudden demand. The Marquis de Sade mentioned them with approval, though one wonders if he was thinking of the then-common sheep's-intestine type, or the porcelain or otherwise modified types from the Orient. Casanova recounts in his memoirs how he arranged to "enter into relationships" with a respected woman and at the last moment decided to buy a condom from her to guard against possible infection. She refused to sell them at what he thought was a reasonable price, and then when he agreed made him take a full dozen. Finding that an affront to his ego, he disappointed the lady by shifting his attention elsewhere and using all 12 on her 15-year-old servant girl. Or so the 51-year-old Casanova said.

The condom and other contraceptives were used in a sexist way. So, even, were the alternatives. It was a common belief that if no other contraception were being used, children could be avoided if the woman lay perfectly still during the whole proceedings, evidencing no emotion while the man went about his labors. From this pleasure-destroying belief there was only one brief interlude, in England, for a few years early in the 1700s. A popular manual came out insisting that if a woman *enjoyed* having sex she would not get pregnant. The author gave an example for British couples to follow: "Sometimes the woman conquers, as is the custom of Spanish women, who move their whole body while they have intercourse, from an excess of voluptuousness (they are extraordinarily passionate), and some of them passionately sing a song, and on account of this Spanish women are sterile." The staid town clerks and civil servants of Georgian England seem to have ignored the information.

Effective birth control, especially types acceptable for women, became popularized only well into this century. The first office offering free birth control advice, and instructions on the use of the protective diaphragm, was opened in London shortly after World War I, when British women had just left their houses in hordes to enter wartime factories. Similar clinics opened in the United States, though often persecuted by police until well after World War II. The effectiveness of the scientific methods they encouraged can be seen from the statistic that while the first visitors to these clinics had had an average of four pregnancies before they were thirty, now less than 14 percent of American women have four pregnancies in their whole adult lives.

The first of the modern crop of worthwhile contraceptives was the diaphragm. The name is Greek for "barrier," and that is just what it is. Slipped in over the cervix, it keeps any trespassing sperm from crossing along up to the uterus. Although the geography is impeccable, its psychology is less so. Diaphragm users still have a 17 percent average long-term pregnancy rate, due in large part to the frequency with which the device gets left resting lonesomely in the drawer. The intrauterine device, or IUD, has only a 5 percent long-term failure rate. Its pedigree is long, dating back to Bedouin tribesmen, who put peach pits in the uteruses of female camels which could not afford to get pregnant during cross-desert treks. In more delicate form the IUD was introduced as a coin-sized silver spring in the 1930s, a construction switched to copper or plastic in the late 1950s.

No one quite knows how it works. One idea is that it causes extra contractions of the uterus and fallopian tubes, not enough to hurt, usually, but just enough to pull a fertilized egg along before it's had its three-day growth period in the fallopian tubes, and so is not yet capable of latching onto the uterus wall. Another theory is that it irritates the uterus enough to dislodge any implanted egg — producing a series of mini-miscarriages, noticed as the slightly heavier periods sometimes arising after use of an IUD. The IUD might even call out swarms of attacking cells, that die all around it and, in doing so, release a chemical that is toxic to sperm. Molded in plastic, IUD shapes are limited only by the designers' imagination, whence the Birnberg bow, Margulies spiral, Lippes loop, and Dalkon shield. One twisted in the owner's monogram would possibly work just as well, providing thus perhaps the ultimate in discreet-display status symbols.

Then there's the pill. Effectiveness rate: up to 99.6 percent. Once swallowed, it works by floating in the bloodstream up to the brain, there to fake the pituitary and hypothalamus glands into believing that it is registering the blood of a pregnant body, and lead to the pituitary's sending out another hormone messenger that goes back down the bloodstream into the ovaries to keep them from producing an egg. The path is simple, but the dose is a question. When the idea first came up in the early 1950s, test trials on humans were decided on as being the quickest way to see what level of hormone would work best. The fact that the researcher who did the main checking was partially financed by the Searle pharmaceutical company should not be taken as having influenced his conclusions calling for the hasty test.

The ultimate market was likely to be North America, but because American women would probably not have taken kindly to having a hitherto untried hormone extract tried out on them wholesale, the researchers took their product to Puerto Rico, where in 1957, the governor-general gave his approval. The head of the London Ciba Foundation later called it "one of the most risky and foolhardy measures ever to be put into general use, particularly against an underdeveloped people such as the Puerto Ricans." The test showed that the pill could work, and also produced "findings" that led to cutting its dose by 50 percent. The "findings" were cases of Puerto Rican women's coming down with high blood pressure, brain clots or terrible cramps when the first proposed level was found to be too high. Because of this fine tuning in the Caribbean, the pill's complications today are statistically tiny, around 16 times less than the health complications from childbirth.

The Catholic Church's position against the pill is based on a pronouncement of Pope Paul VI in 1968. The Pope — then 71 years old — was acting against the recommendation of a council of his bishops, who had come out in favor of the pill on the grounds that it did all its work only *before* any egg was fertilized, and so had no relation to abortion. But in respect for papal infallibility (itself a doctrine announced by Pope Pius IX only in 1870 after he assembled a carefully rigged high-level religious council to grant it), most of the Church hierarchy closed ranks behind his choice. Despite the decision, Catholic regions in Europe of a given industrial level tend to have the same birth rate as their non-Catholic neighbors.

For future contraceptives there are possible injected super-pills, that last a year, while from the past there are still numerous widespread holdovers. Perhaps the two most common are the rhythm method, whose users are statistically highly likely to become parents, and coitus interruptus, the practice of male early withdrawal which, one can gather from Biblical references, was in wide use even in ancient times. The fact that the world's population has increased over 4,000 percent since then is perhaps a mark of its limited efficiency. Continued breast-feeding is possibly even worse as a contraceptive: 50 percent of women start ovulating again three months after giving birth; for nursing mothers the delay till 50 percent of them start ovulating again is just another six to seven weeks.

Fetus and Embryo

Now back to the womb. We left the fertilized egg dropping onto the uterus and then starting to consume its way into the nearby submerged blood vessels. In seventy hours of consumption it sinks lower and lower, until it has submerged itself in a little burrow in the uterine wall. That's when the big change happens. So far it's not been clear which parts of it (the fertilized egg) are living, and which are not. Nine days after conception, the inner section starts to turn into the fetus, and all the rest develops into the less animate life-support system. It is helpful to start with the latter.

Surroundings

The shiny outermost curve of the fertilized egg is what develops into the embryonic sac. That's where the fetus will float, and so that's where the rest of the life-support system must lead. The simplest arrangement would be to have the developing fetus continue to take nutrients and oxygen directly from the mother's blood circulation. After all, the first thing the fertilized egg did when it landed on the uterus wall was burrow down precisely to reach those nutritious blood vessels. But there's a problem with that, and because of it a new system is needed after the first few days.

The problem is that the mother's blood supply is just too powerful to go directly into a tiny developing fetus. In her blood system all the sex hormones of a full-grown woman are constantly swirling, and it wouldn't do to have them rush directly into the growing cells of a fetus. Even if the fetus is female those adult hormones would be too strong. Also fluctuations in the mother's habits — lots of squidgy fat after a heavy meal, storming rushes of adrenalin during a tense taxi ride — would produce changes in blood composition too strong for the little one to handle. The new system that gets around this problem is contained in the placenta. This is an organ that developed out of the first cells the newly burrowing embryo sent into the uterine walls and continues to rest on the dividing line between mother and child. What it does is to feed carefully graded food and oxygen into the amniotic sac in which the fetus is resting. That way, all sudden, overpowering surges are avoided.

The placenta works in a remarkably simple way: it has two sides — a bottom that winds its way in and out among the blood vessels that lead from the mother into the uterus, and a top that narrows into a channel leading toward the fetus. Overall it looks like a weirdly distorted plate, with the bottom shaggy and scraggly (to reach all round the mother's blood vessels) and the top smooth and clear, with a single

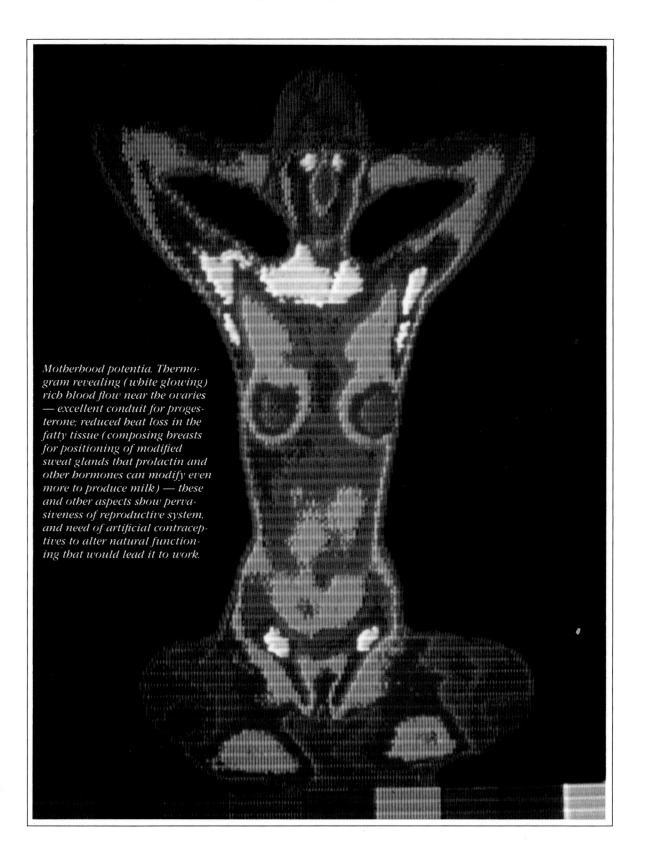

Motherhood potentia. Thermogram revealing (white glowing) rich blood flow near the ovaries — excellent conduit for progesterone; reduced heat loss in the fatty tissue (composing breasts for positioning of modified sweat glands that prolactin and other hormones can modify even more to produce milk) — these and other aspects show pervasiveness of reproductive system, and need of artificial contraceptives to alter natural functioning that would lead it to work.

lead stretching up from it to feed into the fetus. The smooth top is a circle seven inches across, and has a surface area equal to a 45 rpm record, or a small flat cake. (*Placenta* is Latin for "cake.") The bottom, with all its stretchings and bendings, reaches out for a much greater total surface area, and would add up to 140 square *feet* if stretched flat — as much as a good-sized room.

For all its great size, most of the placenta bottom never actually touches a single one of the mother's blood vessels, for that would let in the direct blood flow it is there to try to avoid. Instead, its scraggly bottom creeps in close to the tiniest blood vessels, and then just sort of hangs around there patiently waiting. The wait is not for long. For the placenta itself leads directly out from the fetus, and so is filled with the blood from the fetus. That's where it gets its power. The fetus's blood could almost be termed superblood. When it flows down toward the bottom of the placenta and scuttles to one of these outlying scrags just a fraction of an inch away from the mother's bloodstream, there is no contest at all. The far more attractive fetal blood makes the oxygen, vitamins, fats and dissolved carbohydrates in the mother's blood leap out from her capillaries and shoot across the gap to the welcoming fetal blood in the placenta's outcroppings. The adrenaline and sex hormones in her blood get tugged a bit, too, but not quite enough to fling across. They stay safely out. Over each of the 140 square feet it's the same story, and that is how the fetus gets the food and air it needs to live and nothing unwanted along with it. Useful anti-infection agents — antibodies, bacteria dissolvers and whatnot — also make it across from the mother. Alcohol, carbon monoxide from cigarette smoking, and many drugs are unfortunate exceptions, and do get through.

But how does the freshly bolstered blood, safely uncontaminated by direct touch with the mother, get into the fetus itself? There is a cord to carry it. It stretches from the placenta on the surface of the uterus all the way up to the belly of the fetus, reaching it precisely at the point where its removal later will create the navel. As "navel" is *umbilicus* in Latin (and they saw it first), the structure is called the umbilical cord today. It is made up of one hollow vein for the maternally loaded fresh blood, and two exit arteries for the fetus's ensuing used blood. The three gurgling channels slot into individual sheaths. They're coated with a jellylike substance, surrounded with a paper-thin skin, and then coiled around each other in what looks like a free-floating rope, dangerously close to the unsuspecting child-to-be.

The average umbilical cord is 20 inches long, though ranges from 5 inches to 5 feet have been measured, and it's easy to think that in all that length there's plenty to get entangled around the often gymnastic embryo. But the blood pumped through it by the fetus's heart travels at a good 4 mph, which stiffens most cords like a garden hose being used, so making them taut and noncoiling. Even when it does slip casually around the embryo's head it's likely to be long enough to just as casually slip off. The cord is so long compared to the embryo that there have been safe births when it was looped four, five or even six times around the embryo's neck, each loop being loose enough to

Sugar processed by the liver into glucose: the form in which the body, and the embryo within the body, need to use it.

cause no damage. Modern parents delivering a baby at home have a good chance of getting a shock from seeing their offspring come out with the cord wound around its neck, a sight midwives and doctors have become used to, since it occurs without incident in over 20 percent of all births. Though it can sometimes catch fatally, late in labor when the baby's head is in the pelvis — an event of which careful monitoring of fetal heartbeat can provide early warning.

To round out the intrauterine geography, there is the amniotic fluid. *Amnion* can mean "little lamb" in Greek, and that's fitting here for that pleasing creature is often born with its translucent amniotic sac still wrapped neatly around it. The amniotic fluid is the stuff the embryo floats in, there in its appropriately labeled amniotic sac, and it always conjures up an image of total bliss and relaxation. All attempts to reproduce that fully surrounded floating, be it beach, hot bath, or heated Jacuzzi, are doomed to failure. The superiority of the amniotic fluid is that it not only surrounds the embryo, but it also is swallowed and "breathed" by the embryo. Because it's inside as well as out, it provides a totally relaxed sense of buoyancy. Even scuba divers don't breathe the water they float in and therefore can get pains in their chests and throat walls from the high-pressure air within them trying to push out. But for the embryo, the amniotic fluid has equalized pres-

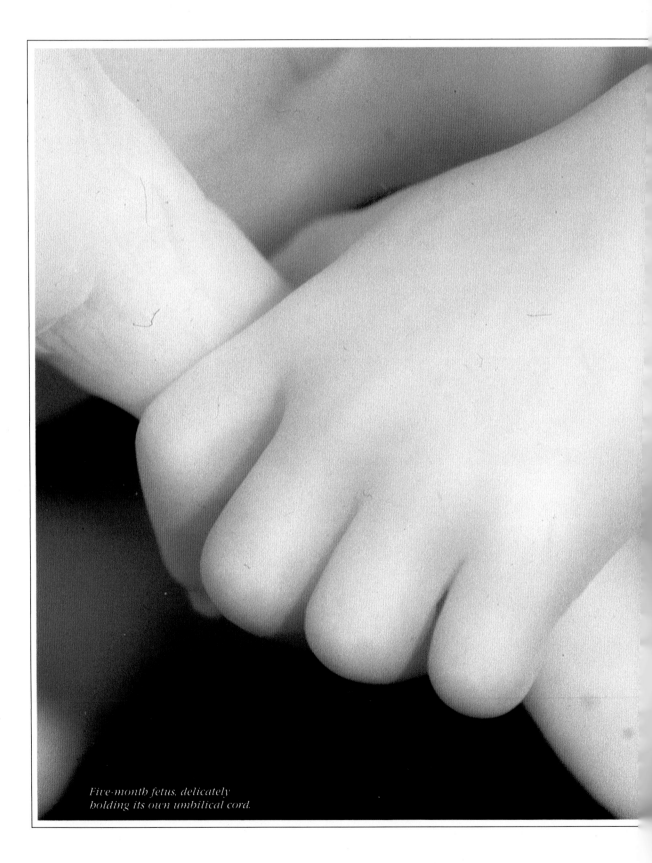

*Five-month fetus, delicately
holding its own umbilical cord.*

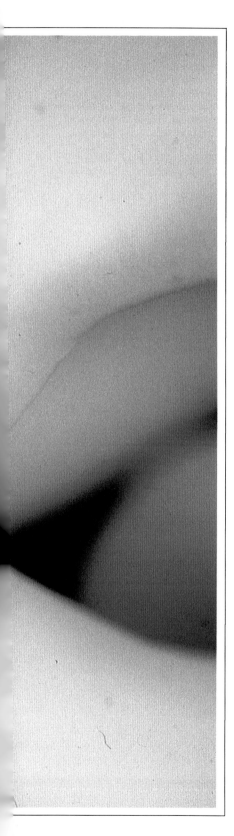

sure inside and out, putting no strain on the chest or anywhere. And since the embryo is being supplied with all its food and air directly through the artery leading from the placenta, there's no problem with having the fluid filling up the lungs — they're not used until birth.

Without this fluid inside it, the embryo would not be able to do anything other than lie flat down on the bottom of the uterus, for it doesn't have the muscle or bones needed to stand against the force of gravity. This is clear once it has to give up the fluid at birth, for no newborn, not even the future Arnold Schwartzeneggers of this world, is able to stand or even sit up at birth, let alone do the gymnastics that the mother can so easily feel it carrying out in the womb. Comforting and supporting, amniotic fluid also provides a convenient waste-disposal system for the developing embryo's certainly quite active bladder. Urine sent into the fluid gets diluted to a nonirritating level and routed back to the mother's body for disposal by the embryo's swallowing it, going into the umbilical cord, and from there out the placenta and into the mother's blood, kidney and ultimately, bladder.

Most important of all, the fluid an embryo subsists in makes for an ideal protector from shock. Any hard blow on the mother's abdomen will end up compressing the fluid, and not the embryo within. Mouse fetuses have been able to withstand accelerations of 3,000 times the force of gravity this way, which compares favorably with the maximum of 22 times the force of gravity that even the fittest astronauts can take. (Dogs have been made to survive half an hour breathing properly oxygenated water, and could survive terrific accelerations without a stop during it, but the experiments never worked beyond half an hour, and astronauts were never given screw-on aquariums to replace their helmets.) The three most frequent causes of abdomen banging in pregnancy are slipping downstairs, on a carpet or the sidewalk and car accidents. They might hurt the mother, but they almost never have an effect on the embryo, who sits them out, floating peacefully away from the shocks, safe and weightless in its warm womb. This soothing, buoyant, and protective amniotic fluid reaches a peak of just about one quart shortly before birth, weighing then some two pounds.

Fetus Growth

So much for the surroundings. The real interest is on the developing person at the center of all these inner ministrations. Two names are used, and in both cases the scientist matches the poet. For the first two months the name "embryo" is used, a gentle euphemism related to the Greek verb form "to be full of life." After that the life form is known by its effects rather than origin, and the nomenclature switches to "fetus" — a term that comes from the Latin for "fruitful" and is a reminder of its recent "fertilization." The division is a sound one, as until two months the growing life is continually changing and building up its basic parts; after that, for the final two-thirds of the nine-month cycle, it's just nourishing and getting larger, pleasingly noticeable to the occupant, perhaps pleasing but sometimes a bit too noticeable to the occupied. The nine months produce an amazing change from the minuscule implanted egg to the wide-eyed, cry-blaring newborn, and a

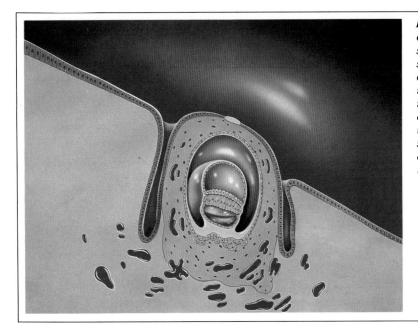

Fertilized egg freshly implanted on wall of uterus; drawn to show beginning of division inside into three layers from which outer structures including brain, middle structures including muscle, and inner structures including stomach ultimately develop. Blood globules at bottom suggest spreading of implanted egg into wall in search of nourishment.

close-up week-by-week calendar will show how it happens. Since the first eight weeks are the most important time, they are the ones that will get the most detailed description.

WEEKS ONE AND TWO This is when the egg pulls in a sperm, starts floating down the fallopian tube, and becomes implanted in the uterus. The fetus-to-be, 1/40 of an inch and weighing less than a dandelion puff, at first blends indistinguishably into the cells that will end up as umbilical cord placenta. In the middle is a hollow, and that is where it slowly comes to life. A smooth disk begins to grow in that space, and as it grows it slips into three distinct layers like the levels in a triple-decker cake. (The disk is balanced on a dimple from the surrounding egg wall, and it is from that dimple perch that the umbilical cord later forms.)

Raised in the center of the hollow egg, the three layers of the disk begin to go their ever so different ways. From the topmost will sprout the skin, brain, nerves, tooth enamel, eye lens, and nose lining of the person to be. From the middle of those three layers later come the blood vessels, muscles, and bones. And from the bottom layer, which ends up tucked in the center as the other two curl around it, will come the internal organs of stomach and lungs, liver and intestines. What exactly controls these changes is not known, as we'll see in a later section.

WEEK THREE Something like a living outline first appears. The fetus is pear-shaped, and within it the most important cavities are starting. There is a hollow tube stretching its whole length that will become the spinal cord, a space for the brain on top, and beginnings of the chest and abdomen cavities to the side. A bulge is forming where the head will develop. By the end of the week there are small pits where the ears and eyes will start up, and tiny swellings that will become the mouth and jaw. The body has two clear sides, which are mirror images of each other. This is a distinction we keep all our lives, but for a few exceptions such as the thighbone and lung, which are slightly larger on the right side, and the testicle and breast, which are equally a bit bigger on our left.

A memento from this time when the two sides first formed, one that we all have, is the slightly dark line that can be seen stretching down from navel to genitals and is especially clear in dark-haired people. This isn't the only leftover seam we have from our fetal days: we'll meet the others later. The three week fetus is 1/10 inch long, and weighs but a fraction of an ounce. It has held up the mother's period by one week now, and the progesterone from her ovary that's supporting it might be making her feel a bit queasy.

WEEK FOUR Now things really pick up. The open tube along the back starts closing over in the middle, and continues closing up all the way along. The top section swells into the rudiments of the brain, though there's still an open hole at the very tip where the closing-over hasn't been finished. The sheeting over this tube at the bottom forms what looks like a little tail. Buds that will turn into arms appear, popping out from nothing in just two days. They're followed at the end of the week by two lower buds that will stretch into legs. This lagging of the lower body behind the upper continues all through pregnancy, and even up to the infant's third year. A small tube in the middle of the fetus curves into a U and starts to squeeze open and closed at irregular intervals. This is the heart. In proportion to the body it's nine times larger than the adult heart will be. The first blood cells for it to push are formed, and a leftover yolk sac that was in the egg, and is still the size of the whole fetus, turns into a red-cell-making factory. From that yolk's wall, the first primitive sex cells have migrated over to what will be the fetus's sexual organs.

That's not all for this extraordinary fourth week. While the tube at the back of the fetus is turning into the spinal cord and brain, another tube in the middle is formed and starts developing into the intestinal tract. From its top are formed two lungs, at this stage tiny and still very solid. The kidneys appear, though they are not yet able to do much of anything and with just the uneven squeakings of the U-shaped heart to drive the circulation, they wouldn't have much blood to clean even if they did work. Around the side twelve streams of cells begin to flow and gradually harden into what will be the ribs. Where the streams jam together at the front is where the breastbone is formed!

The head and neck are half the whole body length, compared with ⅛ the total length in the adult, and they loll so far forward that they almost touch the tail. The ear pit deepens for where the ear canal and balance centers will appear. There are nostrils but no nose, eyes beginning but no eyelids, small cheeks, and dangling from the middle of the head, a clearly started tongue. Just below the head, looking for all the world like a row of double chins, are a series of six ridges resembling the gill slits of a fish.

At the end of the week, one month since conception, the fetus is ¼ inch long, and has undergone the fastest growth it ever will, now enlarged 10,000 times from the original fertilized egg. Pregnancy tests at this point give clear results. The fetus still looks vaguely like some sort of insect larva or fish; four weeks later it will look like a man.

WEEK FIVE The head is still almost half the total length, but it is beginning to straighten up from its tucked down position. In it the brain now has two rounded cerebral hemispheres, and behind it more of the original back-cord is being squished up to form the midbrain and cerebellum. External ears can be seen, as well as black indentations for the eyes. The upper jaw and lip form by two streams of material sliding together from the cheeks. Where they meet, the upper lip is likely to show a ridge, which most of us still have. The lower jaw forms in the same way (its remnant being the less common cleft chin — à la Kirk Douglas) and between them the mouth appears. External ears and nose begin. The heart, still a tube, is now beating at a regular 65 strokes per minute, sending the fetus's own blood through a well-developed network of arteries and nerves. A few nerves begin stretching down from the brain and out from the spinal cord. Because of the head and neck's great length the heart is almost literally up in the throat, and the nerves to control it slither out from the brain to the throat accordingly. As the heart migrates later down across the shoulder to the chest, it unfurls the nerve behind it. This is another peculiar holdover to adult life, and means that you are likely to first feel a heart attack in your shoulder, and that's where the nerve still stretches.

The lungs are bigger than before, but still solid as a hard rubber handball. The arm buds sprout into shoulder, elbow and hand, followed two days later by five buds at the end of the fingers. With their usual delay the leg buds follow suit, turning into hips, knees and toes in turn. The stomach appears. The length is now a half-inch.

WEEK SIX Growth continues to be fastest on top. The two cerebral lobes shine through the still-translucent skull, and nerve connections within it are linking up at the rate of 250,000 or more each minute. The fresh brain cells creep forward into place like an amoeba squishing its pseudopods and manage only a fraction of a millimeter's travel each hour. No more than 3 percent of the brain cells get lost and hook up to the wrong parts of the brain in all this blind spaghetti weaving; luckily the wrong ones are likely to melt away before birth.

By this time the tip of the nose has come out, the mouth is taking on its final form and another most striking growth is in the eyes. These began as two droopy channels lifting out from the brain's surface, one going to each side of the head. When they reached the surface, they folded down into smoothly polished cups. At the point where they break the surface, the skin starts to stretch over the gaping opening into that cup, flashes clear like glass, and thickens into a perfectly shaped lens and cornea. A million nerve fibers will grow from the back of the eye and stretch in perfect formation to a visual switching station in the middle of the brain, before the fetus's growth is done.

Human fetus — bones stained to show skull bones not yet joined on top; bones of the eye socket and around the mouth are still developing too.

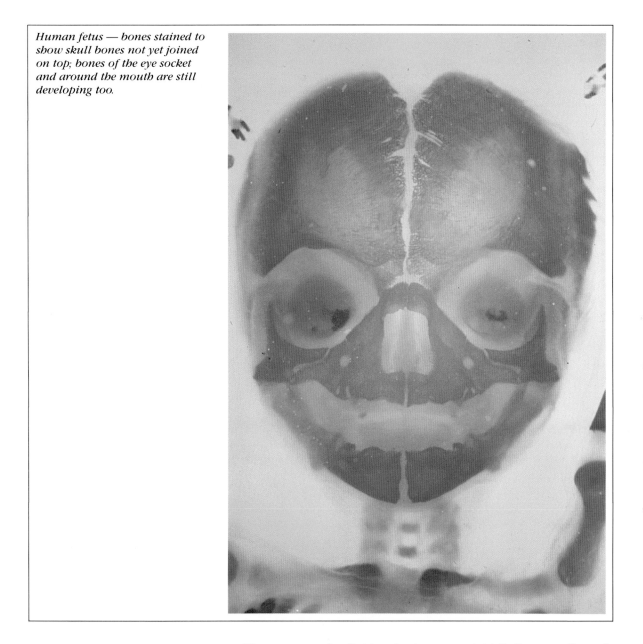

The outer ear also finishes, having grown initially from two flaps of skin. One curved up from the back to make the shell of the ear and the earlobe, while the other swept up from the front, to cover the canal leading to the eardrum. Where they meet, at the top of the external ear, there's a slow-motion crunch, and the leftover from that collision is the little nick we all can feel up at the top of our ears.

The skeleton starts showing up, though first it is made out of cartilage instead of bone. Blood cells will have to continue coming from the yolk until the bone marrow and liver have been formed. The fingers grow out clearly from the arms, minute but perfectly clear protrusions on an arm that itself is still less than the size of an apostrophe (').

The fingers and thumb are all the same, like the oversmooth outline of a cartoonist's sketch. The arms and hands stick straight out of the body, unable to touch because of rigidly locked elbows. The whole appearance is rounder and more soft than the week before, fleshed out with clear muscle bulges at shoulder, thigh and neck. The little dimple in from the placenta on which the fetus originally balanced, has almost completed its change into the umbilical cord. Boys and girls start their separate ways. Special sex cells swoop over to the kidneys from their scattered position throughout the fetus, and then, depending on the genetic instruction and circulating hormones, start to form testes or ovaries. The fetus is ¾ inch long, weighs 1¼ ounces, and can even wriggle a bit, though not enough for the mother to feel.

WEEK SEVEN Enough nerves have arced out from the brain and spine for the fetus to have its first reflexes. Stroked on the upper lip, it will show the basic responses of the classic adult startle response: back pulled in, arms hurled out, and the whole body trying to twist to the side. The mouth has lips and evenly spaced buds for the twenty milk teeth. Internal organs are completed, with the blood vessel network now linking up the whole body, the stomach secreting digestive acid and the liver, so big it bulges temporarily into the umbilical cord, making some blood cells. The arms are no longer stubby but slender, and they almost reach each other. On their ends the thumbs are growing differently from the fingers and on the feet the heels and toes are clear.

Family differences even begin. They show up at first in obscure places, some fetuses having big earlobes, other little ones; some tapered waists, others broad ones. The length is one inch and weight two ounces. At the same seven-week gestation time a blue whale fetus weighs 140 pounds and is as long as a small cruise missile, while a rabbit fetus is no longer a fetus at all, but quite likely a great-grandparent, with the total weight of all its 216 possible descendants chalking up over 300 pounds.

WEEK EIGHT It definitely looks like a human now. The head is rounded and the eyes have shifted forward from their initial post out on the side. The nose looks a bit snub, as it waits for the underlying bone to lift up, but that's more than made up for by the disappearance, finally, of the protruding tail. Most of it shrinks up into the spinal cord, most of the rest gets dwarfed by contrast with the proudly developing buttocks and the little that's left is small enough to fit under the skin, where it stays as our tailbone (coccyx). The fetus is a bit over two inches long, just the length of a little finger and weighs half an ounce, a third of what a small doorkey does.

Importance of Fetal Growth

Eight weeks is the dividing point. Usually it's only if something goes wrong before this time, be it a misread DNA instruction or a dangerous chemical from outside, that there will turn out what parents perhaps most dread, a child that has been grossly malformed. The fear is

Wonder in the Womb — an eight-week embryo. The taut bubble of the amniotic sac is produced by pressure from the fluid inside it, which the embryo floats in and also swallows. Note the tight coils of the umbilical cord, the dark spot by the future navel which is an enlarged liver for the first blood cell production, and the slight depression on the roof of the head outlining how the left and right cerebral hemispheres of the brain are separately developing. Note also the much greater development of the top part of the body than the entire region below the waist, an imbalance only fully made up after a child's second birthday.

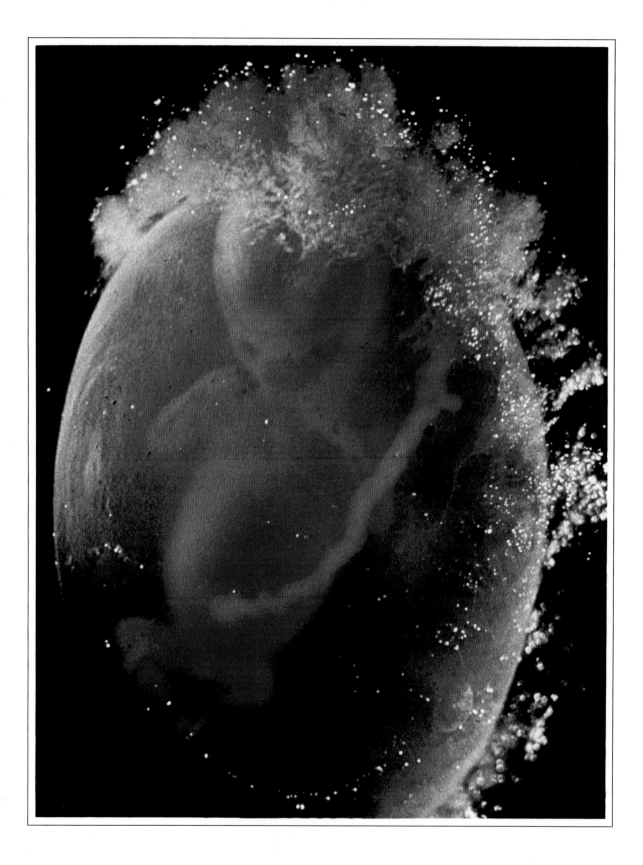

rarely broached, but still it's there, so before going any farther, some reassuring statistics.

When something goes wrong, very wrong, early in pregnancy, the mother's body is going to know about it. (One self-regulated cause, for example, can be a malfunctioning placenta, which is supposed to start producing progesterone, and if it has something very wrong with it, it won't produce enough to maintain the pregnancy.) The result is almost certainly going to be a spontaneous abortion, a closing tight of the uterine surface just under the implanted egg or fetus, and a reflexive flinging of it out of the body. If this happens early enough, it will appear to be only an unusually heavy period that month. It occurs in 10 percent of all pregnancies, or about 500,000 times a year in the U.S.

Everyone who is born has passed this selection point of the body's own rejection. Because of spontaneous abortion the rate of what are coolly called "severe congenital abnormalities" is no more than one per hundred. This means that close to 99 percent of deliveries will produce normal infants. That's a lot. There are also increasingly accurate ways to check if something might be going very wrong before birth that the uterus's reflexive expulsion somehow missed. Samples of the amniotic fluid around the fetus can be drawn off by a thin hypodermic inserted through the mother's abdomen. Done right, it doesn't hurt. (Though as there are side risks it shouldn't be a matter of routine.) In that sample will be a few skin cells that flaked off from the fetus, and with just one of those cells the biochemists can get down to work. As we saw in the genetic section, each cell in your body has on store in it all the DNA information for every other cell in your body. That holds true for a fetus, too. The sampled cells can be treated with the right chemical tools so that their nuclei break up and the chromosomes inside can be counted. A given excess can be a sign of mongolism (Down's syndrome) or other problems. Parents informed of the very worst faults early can act accordingly.

The best hospitals also have high-frequency sonar units, which are so finely focused by computers that they can be aimed into the womb with no possible harm, to measure precisely the rate at which all parts of the embryo are growing. (The sonar unit of a nuclear submarine, although it works on the same principle, is so powerful that it is dangerous for divers to swim in front of it when it's switched on.) Contrasting what shows up here with what's supposed to happen can also tell a lot. Normal embryos undergo head-neck deceleration at a certain stage of their growth. This means the head and neck stop enlarging as fast as they had been at first, letting the rest of the body catch up. Let this deceleration be two days late and a diagnosis of hydrocephalus can be made. This means the head is going to be way too large and the brain will be profoundly distorted. The condition is invariably fatal in the first few days after birth. (Many couples find it easy to choose abortion when the diagnosis is made early enough.) An unfortunate surprise is rare, and getting rarer.

Still, there are many distortions that can't yet be noticed before birth. Almost all of these also take place in the first eight weeks. The example of a growing arm will show why. Remember that it starts as a

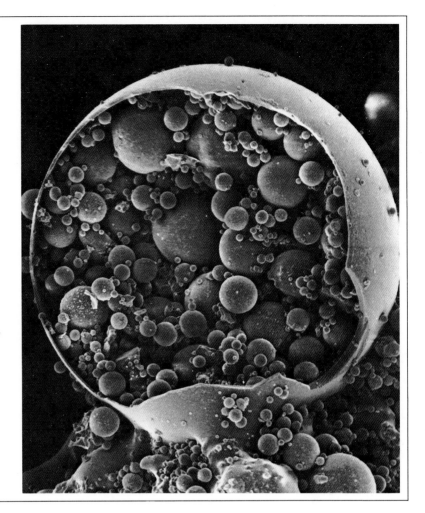

Chemical attack. The proper growth of the fetus is continually under threat by external radiation, ingested chemicals, and atmospheric pollution. This is fly ash, a common atmospheric pollutant, breathed in in the millions by city-dwellers every day, and containing magnesium, iron, glass, and sodium. Though not known to be a producer of deformations itself, fly ash with its extraordinary spheres-within-spheres construction serves as reminder of how many potentially dangerous chemicals are out there and unknowingly absorbed. The whole is less than .004 inch across — the outer sphere gets filled with smaller ones because of the unique temperature and pressure conditions in the high-temperature power-station chimneys where it is produced.

bud on the shoulder in the third week, lengthens to include an elbow and wrist in the fifth, and grows fingers on that base later in the sixth. If anything puts off development in the fifth week, even for a few hours, the fetus will skip the phase of lengthening its arm. Nothing will happen when the wrist and elbow should be formed from the shoulder buds, and since the signal for it only comes once, the arm will never get a chance to reach its normal length again. Instead, when the signal to produce fingers comes in the next week, the fetus will grow them directly on the still-tiny lump on the shoulder.

This abnormality called phocomelia, is usually extremely rare. It seems to be one of those conditions that the uterus will naturally eject in the rush of early, and unnoticed, spontaneous abortion at the fifth week. Recall that the fetus is just half an inch long, and its arms, even if properly formed, are even shorter than an apostrophe. That shows how finely the body can detect early faults. But in the years between 1958 and 1960 phocomelia became something of an epidemic, due to the then-popular drug thalidomide, which had been marketed as a sedative and which many pregnant women took to prevent vomiting. It was for sale chiefly in Germany, starting in 1957, though it was often used by families of NATO soldiers stationed there, and sometimes even taken back home to the U.S. because it was so effective.

Thalidomide has an especially light molecule, which made it work well as an antinausea drug. But this molecule was so light that it could also make the leap from the mother's bloodstream into the placenta, a leap that was normally restricted to life-enhancing ingredients such as proteins, sugar, dissolved oxygen, or minerals. Once in, the thalidomide wasted no time. From the far end of the placenta it took under thirty seconds to scamper through the umbilical cord, reach the fetus and latch onto cells being readied to grow the arms to full proportion in the fifth week. It stopped them from doing any such thing. It also stopped the uterus from rejecting it as abnormal. The results were thousands of children with stunted or nonexistent limbs a few months later. An epidemic spared the United States because of a few conscientious technicians at the Food and Drug Administration, who, despite pressure from the drug industry when trials started, refused to release the new sedative until they were convinced it was safe. Their fortuitous obstinacy might be remembered the next time pharmaceutical companies complain about the time-consuming methodicalness of new drug approval in the United States.

German measles does its damage in the same way. It too holds back certain cells that should be doing their work early on. Only this time it's the ones producing the ears and the inner chamber of the heart. Everything goes on in the embryo at the usual rate after its unerring attack, with the result being a child who has only a small hole where its hearing centers should be, or a hole, gaping or otherwise inside its heart. For this reason one should never, ever, go near a woman you think might be in early pregnancy if you have German measles. The deformity rate is six *thousand* percent higher than usual when the mother catches German measles in her first month.

Later on, of course, it won't matter at all. The deformation rate for births where the mother got German measles in her fifth month is no higher than normal, for then, with the ears and heart already long completed, German measles has gone back to being just the usual light flu and funny red rash. The shame is that all these problems are most dangerous at just the time when the mother is least likely to know that she is even pregnant. Chance overtakes precaution, and a child's future is up to every passerby with the wrong bug, or under-checked drug the FDA lets through.

Usually, though, it works. Usually the fetus's millions of cells get formed just right, shifting and stretching, lacing and weaving, in the most convoluted of all ballets. It seems extraordinary that it should ever come out right at all.

Researchers studying the embryo sometimes call it the French flag problem. Given a jumbled crowd of airport mechanics in a big hangar, all of them wearing blue, white or red jumpsuits, what do you have to do to get them to line up in the single color rows that would produce the three-striped design of the French national flag? Putting one guy up on a crosswalk to yell down instructions is cheating, because the fetus's cells have to make do with just the instructions they can carry around with them. There's no equivalent of a guy with a megaphone in the womb, and even the mother can't play the part, for her blood never directly touches any part of the fetus's body — the placenta sees to that.

Worse, to make the French flag problem anywhere near as complex as the one the fetus faces, you'd need a crowd of hundreds of millions of mechanics, each one in a different color jumpsuit and somehow get them to form the design of some granddaddy of all Appalachian quilts. Remember too that brain cells are not just simple electrical cables, but fluid-filled tubes that might send out any one of several dozen different neurotransmitters, and so each need just the right sticky protoplasm and internal power units to drag these neurotransmitters from one part of the cell to another. It's quite a tangle.

How does a fetus ever manage to create and swing about the right cells to turn into you? In the 1920s and 1930s many scientists thought it must be through an "inductor," some all-powerful chemical that each fetal cell sent out to pull its neighbors into the right place. Now it's clear that DNA must be involved. Fetal cells that are going to grow into the heart turn on the part of their DNA blueprint that has the recorded information for building a heart; the ones that become a toe do so by paying attention only to the part of their DNA that outlines how to build a toe. But how the right cells turn on the right DNA, what help they get from growth factors or other cells, remains a mystery.

So far these observations of the first eight weeks have been fact. There's another one that's fiction, but a fiction that has turned out to be almost as important as the truth. Around the fifth week the fetus grew a tail, and back even at the fourth week it had a clear set of gills. It looked a bit like a fish or a baby lizard. Today that's known to be just a developmental oddity. The tail is there only because the spinal cord is longer than the legs at that point. The bridges that look like gills are just staging points for the bones that two weeks later separate to become the middle ear, the edge of the jaw and, in the case of what looks like the second of the six gills, the tiny bone that rests on the bottom of our tongue. (This last is called the hyoid, meaning U-shaped, which it is, and although little known and little noticed, as the tongue bone it is most definitely human, not fish.)

But before these facts were known, one German biologist announced that the developing fetus not only looked a bit like a developing fish or lizard but actually for a brief moment *was* the same as a developing fish or lizard. At the time when he made the announcement in 1872, Darwin's theory of evolution was first becoming popular and people were eager to find a personal confirmation that man had descended from simpler animals. This could be it, the theory took hold.

Darwin protested that all this theorizing was inaccurate, but he was ignored. Oxford and Harvard taught the theory, great scientists of the day came out for it, and even the psychologist Carl Jung based his nutty personality theory on it, neatly making the switch from embryos to dreams by writing that the ideas of prehistoric man live on in our

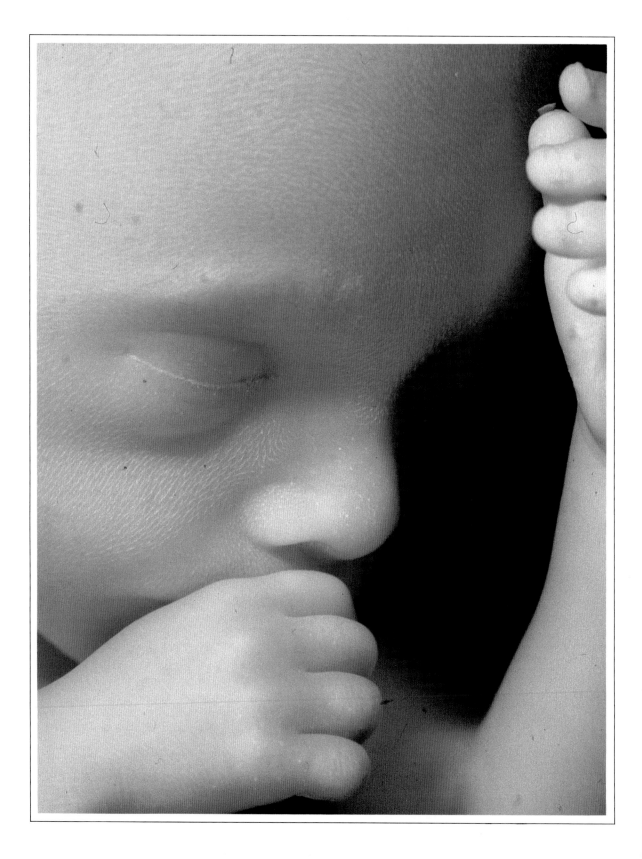

dream life. Dr. Benjamin Spock advertised his belief in the theory in the early editions of *Baby and Child Care,* pointing out that "as he develops [the child] is retracing the whole history of mankind."

The crackpot theory of somatotypes also came from this. The notion that there are three main physical types: pudgy ones who are alleged to have developed from the innermost of the layered disk in the implanted egg, muscular ones who came from the middle layer, and skinny, intellectual ones from the top layer. And if you're one, you can never switch to another. Even though it has been discredited by modern embryology, vast holdovers of this tripartite determinism persist. They linger on in each thirty-year-old career woman who fears that her cleverness means she will never develop the curves or childbearing capacity she would like, and each pasta-bound, overweight businessman who in the back of his mind knows, knows with only the sureness a beguiling error can produce, that no jogging regimen, no diet or fad, can ever change him from his God-given stomach-broad form.

These theories all were wrong. A return to the contemplation of the fetus, contentedly waiting in its little sac, not fish or lizard, not predetermined muscleman or fatty, would have shown why.

Later Pregnancy

If the whole course of pregnancy were fitted into a single nine-to-five working day, all the important development in the fetus would have taken place by the mid-morning coffee break. From then to the end of the day it's just a wait. In those final seven months there's little ground for mystery in the developing fetus. What it does is one thing, and that it does without doubt: grows. The 2½ inches and ½ ounce of the eight-week fetus swells into the 20 inches and 7¼ pounds of the average newborn infant, puffing up the mother as it goes by an extra 14 inches on average around the waist.

Aside from the steady growth, not much is happening in the burgeoning fetus. Where the first eight weeks were loaded with changes every day, the last seven months are marked by only a few occcasional debuts. At 12 weeks the sex organs are clear. Urination starts then, as also, with the greatest ecological efficiency, does swallowing. Another month along, hair, eyelashes, and eyebrows have started, accompanied in most cases by the first fetal hiccup.

The fetus is soon making most of the amniotic fluid itself, squirting it out of its kidneys. This is a major process, for the amniotic fluid, once made, does not float stagnantly around the fetus for the remaining months till birth. Instead, it's continually being sluiced out by the mother — ⅓ or more of it draining into her blood every hour — so the poor little fetus has to lock into high gear to keep its protecting buoyant fluid in place: it gushes out four gallons of it a day, every day, for the second half of pregnancy, just to make up for what the mother sluices away.

Twenty weeks, not yet five months after conception, brings the first chance of survival outside the womb. At 13 inches, and weighing 1¼ pounds, the fetus gets 5 percent odds on surviving a premature birth.

A fetus from the fifth month of pregnancy, peacefully sucking its thumb. The gesture with the left hand is not reflexive, but changes in accordance with the embryo's brain waves. Most features are completely formed — as seen in the presence of right eyebrow and the fingernails on left hand. The remaining four months will be mostly steady enlargement.

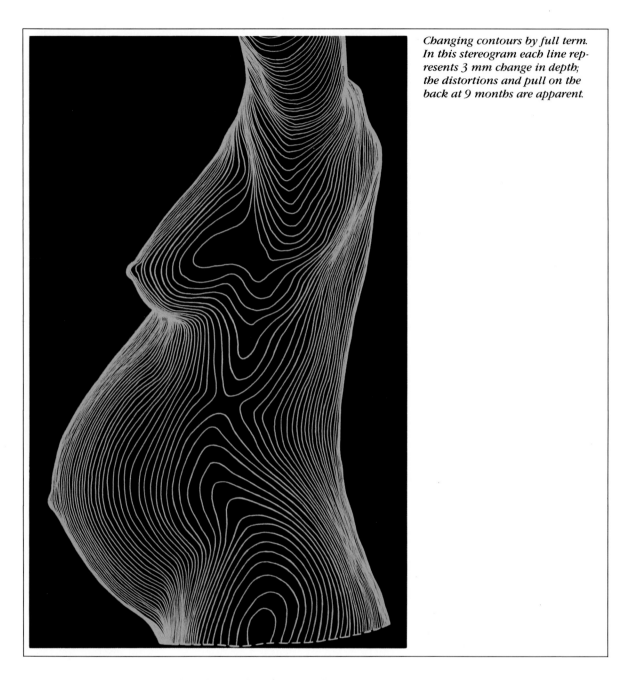

Changing contours by full term. In this stereogram each line represents 3 mm change in depth; the distortions and pull on the back at 9 months are apparent.

A month after that, now weighing better than 2½ pounds, premature survival rates are closer to 15 percent. Two weeks later, still 2½ months before birth, thumb-sucking begins, the embryo gets coated with a special thick grease layer to keep it from getting waterlogged, and the tightly closed eyes will falteringly open if it is born feet first.

The real interest now is with the mother. From the time this eight-week-on stint of steady growth begins, a woman should be able to feel clearly that she's pregnant. The earliest symptoms are lack of a period, nausea, a sharp metallic taste in the mouth or reduced appetite, more

Adult ovary, nine times life size. It contains 300,000 partial eggs at birth, 10,000 at puberty, and is likely to release 400 mature ones in a lifetime. The shape is bulbous to maximize space for the eggs and to allow for the rich blood supply that supports them. The ovary is also a major hormone source: that is, from the crater left by a released egg (a crater which can stretch from one fissure on the photograph to another) a pulpy yellow mass forms that secretes progesterone, crucial for preparing the uterus for pregnancy, and other changes.

frequent urinating, and breast swelling or tenderness. These are all grounds for suspicion, but too common to be sure. Less uncertain are an enlarging uterus, a clear fetal heartbeat, and the first fetal kickings, often described as being like butterflies (or, less kindly, like gas). These appear only well into the pregnancy, at three months or more — thus the frequency of pregnancy tests.

Even ten years ago these tests often involved giving a blood sample to a laboratory, and waiting the two weeks while it was injected into a large toad. Then someone had to stay to see if a glob of tadpole larvae came out. If so, the donor of the sample was pregnant. That test worked because human sex hormones are not much different from the ones flowing in other animals, an aspect of harmony with nature that can be discomfiting upon reflection.

More recent lab equipment can measure the changed hormone level of a pregnant woman directly, so leaving the great toad to its own reproductive machinations. If the test comes out positive, most American women elect to go ahead with the pregnancy to birth. (The therapeutic abortion rate is estimated at around 12 for each hundred live births. That's low internationally, with the figure in West Germany being around 25, in Japan and France closer to 50 and in Hungary a

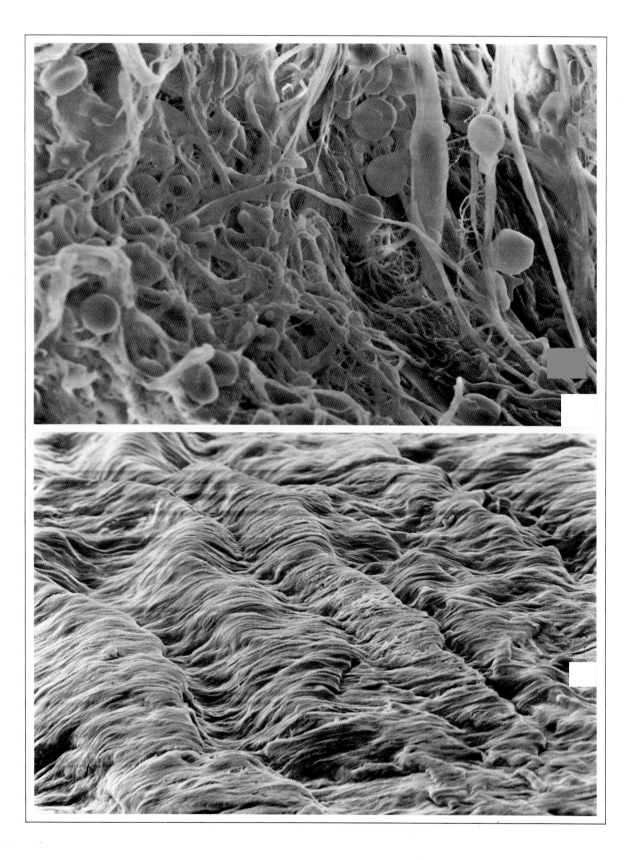

record high of 126 — more abortions than births.) The changes the women start to go through during the long haul are of two kinds.

The first is hormonal. Without some sign to the body that a pregnancy was taking place, all fertilized eggs would just get swept out of the uterus before they had a chance to grow at all. The main check against the overhasty departure is the hormone progesterone (a name meaning simply "supporting pregnancy," progestation). At first it comes, of all places from the little niche the egg left behind in the ovary when it was kicked out at the start of the whole process. That niche grows and grows after the egg leaves, until three months into the pregnancy it's as large as the rest of the ovary altogether. Triggered by signals of a successful implant in the uterus, this pit creates, for the first two months, even more progesterone from the odd chemicals lying about here and there in the ovary. A little carbon here, a little hydrogen there, other precursors of a steroid molecule, and, at a rate of several thousand molecules per day, the niche in the ovary sends the crucial progesterone bubbling off. Each completed molecule looks a bit like two miniaturized versions of the starship *Enterprise* that have collided, having two sets of rings at the center, and various thin bits of debris (but how precise!) suspended down along from it.

Streaming through the bloodstream, the progesterone arrives in the uterus less than a half hour after release. In there its odd structure really comes into its own. The progesterone fits into tiny reception points on cells on the uterine wall, berthing docks really, and by its presence in them forces the uterus to keep on increasing its blood supply to the implanted egg, and, above all, not to start any of those squeezing contractions that at other times would so easily herald the start of menstruation. The uterus follows the progesterone's orders and keeps on building up the thickness of its walls, so that it has enough material to serve as a base for the blood vessels that the young fetus is so hungrily draining.

A few of the progesterone molecules are less well aimed than their partners, however. Instead of heading straight for the uterus, they end up falling through other capillaries into the general blood circulation and tumbling on in there all the way up to the brain, where they help release prolactin. That hormone causes the ends of the twenty or so supporting glands in the breast to change about internally until they're able to modify normal sweat into drinkable milk. Drinkable but not perfect: human milk is a compromise between a dangerous nutritional drain on the mother and an ideal diet for the baby. It's poor in biotin, calcium, riboflavin, and several essential fatty acids.

To keep matters from getting off to too hasty a start, the now freshly active milk glands are kept collapsed in narrow bundles, and even stuffed full of discarded skin cells. Estrogen molecules scooting up from the placenta see to that. The occasional dribbles of milk early in pregnancy are a sign that the blockage has been less successful than the progesterone intended it to be.

This first activation of the milk-sweat glands inside the breast demands a good deal of space, and an uneven rush of prolactin can cause the breasts to swell by up to half a pound in a single day. When that happens the skin stretching over the breast gets pulled sideways faster than it can grow fresh elastic tissue to take up the stress. The result is that thousands of valiant elastic tissues already there, each one smaller than a thread, get torn apart strand by strand. This is the origin of the stretch marks that about 50 percent of pregnant women end up with. Unfortunately, torn elastic tissue never heals, and stretch marks, once created, are never going to go entirely away. The only consolation is that with time they change color from the irritated red produced by the tear into less noticeable dull silver streaks. (Any sudden weight increase will cause stretch marks, which is why fast-growing adolescents are likely to get them around the hips or thighs.)

The breasts are not the only place to harbor hormones that missed the track to the uterus. And hormone molecules on the loose there are aplenty: there can be over 100 times more estrogen molecules floating loose in the blood at the end of pregnancy than usual. Other molecules land on the cordlike ligaments that hold the bones of the pelvis together, and make them a bit soggier than usual, so that they loosen up and let the bones they're holding shift apart farther than usual. It's not a lot, but it can be enough to make the difference several months later between a snugly successful birth and an unsuccessful ordeal. Since the estrogen supply stays high for several months after the pelvic bones have been loosened, excess molecules of it will end up with less desirability on the numerous ligaments that hold the bones of the spinal column in place. Those ligaments unintentionally become softened too, producing an easily swaying back at just the time when a filling uterus is about to put extra strain on that very spine. It makes the sore back of pregnancy a lot sorer than it would be otherwise. Other hormones set up a thick gooey plug in the cervix, just the thing to keep unwanted infections out. With the great number of bacteria in the vagina, this can be a useful precaution.

As a final plumbing effect, leftover progesterone latches onto the tiny muscles around the veins that stretch back to the heart from the lower body. Without the push from those muscles in top shape, returning blood travels more slowly than usual in the veins. This is sometimes the cause of the varicose veins pregnant women can develop, and it is always behind the change in the coloring of the pregnant vagina. With a slower blood flow the usual pale pink will disconcertingly switch to a much darker hue, or even, especially in women of Mediterranean descent, a very pale blue. Even the fetus's placenta barrier adds to the hormonal upsets in the mother. It trickles out lots of enzymes, some cholesterol, even a range of stress hormones.

The other reason for a woman's changing feelings during the final seven-month haul is due to the uterus itself. It's bigger, and it makes its size known by pushing. While an average woman has 17 square feet of skin when not pregnant, she has 18.5 square feet when the uterus is at its nine-month peak. By that time she will have so many extra blood vessels running to supply the placenta and fetus that her blood volume will have gone up by a whopping 25 to 40 percent, three quarts more than usual. They dilute the fixed number of energy-supplying red blood cells she has (although that thinning out alone doesn't make her more tired: increased progesterone does). It also puts a corresponding 25 to 40 percent greater strain on the heart, which translates into a faster pulse. Not just the fetus demands more blood during pregnancy. The woman's intestines take more, her breasts — with their average 1.8 pound growth — each take more, and even her forearm demands more blood, sucking in 2½ times more fresh blood by the end of pregnancy than before.

The steadily enlarging uterus, which rises to the navel by the sixth month and to the bottom of the rib cage during or after the eighth, pushes up on everything it reaches. Estrogens in the mother's blood have made each one of its muscle cells grow: so tiny before pregnancy that it wouldn't have been wide enough inside for even a narrow pencil to fit, now the uterus is broad enough to encase the kicking, punching, pirouetting newborn-to-be. The heart, which normally dangles way up above, gets pushed out of position by this rising behemoth from below. By eight months it will be lying sideways as it beats, caught on top of the swelling uterus. The stomach is also given an unwanted ride up and sideways, being lifted until it's a flabby pouch turned sideways onto the small intestine. There it most unpleasantly can send out a sloshing back-flow to the esophagus, so producing a sharp burning sensation and the general loathing for heavy foods for which pregnant women are noted.

All together, about 12 pounds of the average 28-pound weight gain during pregnancy come from the uterus and its cargo of placenta, fetus and amniotic sac. The remaining sixteen pounds are divided between the breasts, general bodily fluid and stored fat. (The last, although being burned only after all food is used, can be surprisingly easy to get rid of later as progesterone falls.)

How the woman handles these changs is not as dependent on age as usually thought. So many of the changes — progesterone, prolactin and all the rest — are tailored by the fetus itself that she almost always will provide a near-ideal home for the developing cargo. A woman having her first child at 19 has a 97.5 percent chance of delivering a live baby that will survive its first week. A woman two decades older, having her first child at 39, has a 96.3 percent chance of doing the same. The difference is only 1.2 percent.

Since it's so often overestimated, a certain number of women in their late thirties or early forties who take precautions against having a child but in fact desperately want one, will come down with what's called phantom pregnancy. In this sad syndrome most of the signs of pregnancy will occur despite there being no basis for them in the uterus. The mimicking can be extraordinarily hard to differentiate from the real state. Weight will climb on schedule, and the abdomen will slowly enlarge. The woman will feel nauseated in the mornings and will also show the enlarged breasts appropriate to a developing pregnancy.

Also the weight in a pseudopregnancy comes through precisely regulated fat additions, the abdominal swelling is due to intestinal gas, the breasts grow because of a rise in the usual premenstrual hormones, and the regular morning sickness is just a perfectly acted, and quite

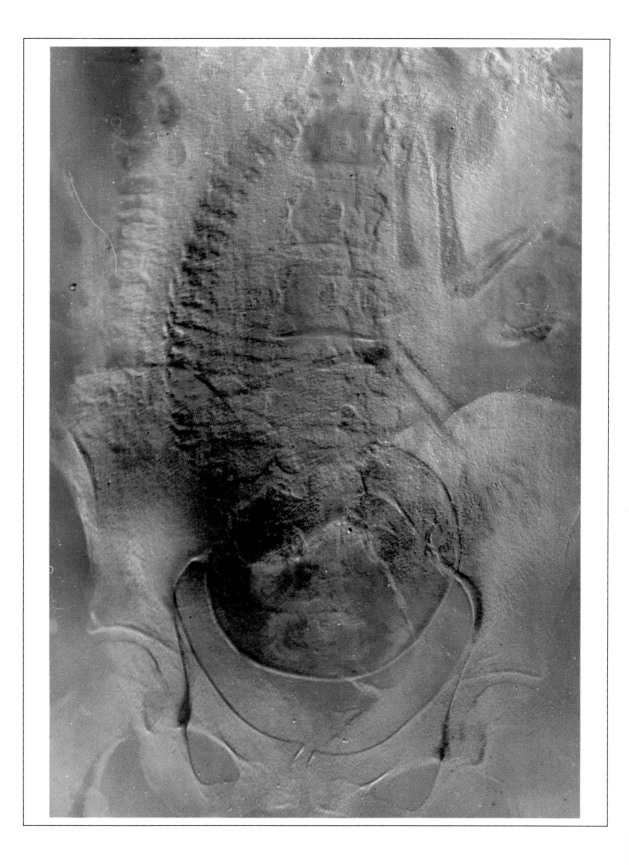

X ray of eight-month fetus in mother's body, just about time when it's flipped over into the head-down position it will keep from now through delivery. Green is mother's spinal cord, pelvic girdle. Orange-brown is fetus: legs at top, curving spine on right, arms at middle, and large, four-inch broad head forming circle between hips. The average eleven-hour delivery is a matter of moving those seven inches through the cervix and vagina and out of the body.

unconscious, imitation of symptoms that have already been seen or read about. The treatment calls for sympathy and understanding.

Birth

Finally, birth. The most common time is early spring, nine months after the summer vacation, followed in the frequency list by a four-week burst in September, nine months after Christmas, and all that festive cheer. The length of the average pregnancy is 266 days (though errant sailors can be bagged in paternity suits so long as their last shore leave was within *398 days of* a legally contested birth). Computing the target date is not helped by the prissy tradition of always inferring it from the time of the last menstruation, rather than asking when the last sex act was and computing it directly. Many culculations of expected birth date are wrong because of a misunderstanding here.

All that really counts is how far the fetus has developed. It does best if born at a certain stage in its growth, and there are only grounds for worry if there's any delay after that point. Staying inside too long, a fetus will start digesting its body fat, swallowing its cushion of amniotic fluid, and quite possibly even gouging itself with its now too-far-grown fingernails. Most worrisome of all, the placenta might begin to deteriorate.

Prematurity, though, is a more common problem than late births. Four to six percent of babies come out before they're fully grown; a rush of hasty arrivers which has included Charles Darwin, Napoleon Bonaparte, the single-named Voltaire and even Sir Isaac Newton, possibly the most intelligent being in history, who was born on Christmas Day, 1642, and put in an open dresser drawer by the fireplace on the off chance that the warmth might help him survive. Those early entrants probably weighed no less than 4½ pounds. Below that and they wouldn't have had a chance. The best incubators today however — enormous, $80,000 devices coated with microprocessors and colored wiring — can sometimes nurse a baby of barely 1½ pounds (that's less than a quart of milk weighs) safely into its normal size. Prematures are more common than late arrivals, because the embryo has a sure-fire way of starting the whole labor business off. The overweighted mother, tiring at the end of her nine-month haul, has no choice but to wait for the baby to act.

What the fetus does, once it has reached its most sturdy size, is to let out a hormone that makes the uterus start contracting and so prepare to send the baby out. Instead of just floating over the two inches or so to the surrounding uterine wall directly, this hormone takes a more indirect path. Fired perhaps from the embryo's own pituitary gland, it goes through the navel and scoots along the umbilical cord. Sometimes it works directly on the uterus while other times it goes up through the mother's circulation until it reaches the brain. There it docks, squiggles things around a bit and after a while makes the brain send the uterus a chemical signal causing contractions.

Curiously enough this will not be the first time the uterus squeezes tight: far from it. Every twenty minutes during the whole pregnancy so far it has been clenching for brief twenty-second bursts, before letting

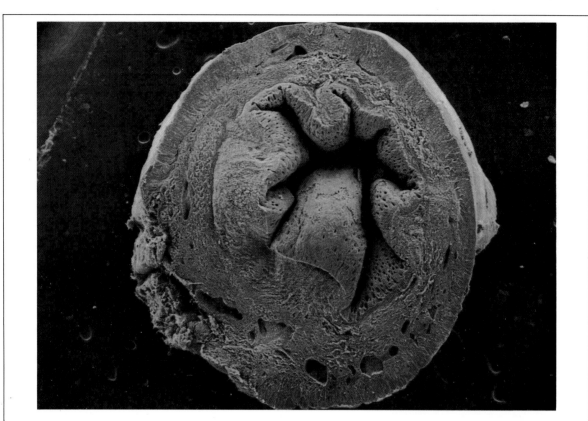

All the power of the muscular uterus is needed during labor to stretch cervix opening (above), here pictured life-size at initial circumference (inner circle) and fully extended (outer circle). This process is the major effort, and the source of the discomfort, in childbirth. The small holes visible in uterus bottom are for blood vessels to nourish its enlarged muscular wall.

go for another twenty-minute rest. That was a way of keeping its muscles in tune and also of helping the placenta squeeze blood into the baby. What's different this time is that the embryo-inspired brain chemical makes the contractions occur a good deal faster, and a good deal harder, than they have done before. When that happens the woman need have little doubt what will come next. The labor has begun.

A recap of the geometry is in order. The baby's in the basketball-sized uterus, which has a thick and strong top part, and a thin and almost flabby bottom part. Under it is the cervix, which leads to the vagina, which itself leads out into the world. The cervix is something like the neck of the uterus (its name is direct Latin for "neck"), but at this stage in the pregnancy it's a very narrow neck indeed. In length it's as long as a cigarette filter, but in width the cavity at its center is no more than 2 mm wide, just about the size of a large dot. Somehow that's going to have to get widened to fit around a baby's four-inch-broad head, and the effort do so is what the major part of labor is about.

Now whatever happens to the cervix so that it enlarges enough to get the baby out, it's got to happen so gently that there won't be enough damage to keep it from happening again, during a possible second birth. All women alive today are the descendants of other women who on the average were able to have at least two children; otherwise the population would have disappeared long ago in the Paleolithic. What's truly significant about a woman's reproductive system is not that it's designed to give birth, but that it's designed to give birth twice. That's why a first labor takes the average fourteen hours it does.

Most of this labor goes into cranking open the cervix, and that has to be done very gently. To see that there's no damage the uterus muscles do not pull directly out on the cervix. That would be awful. The uterus after nine months of pregnancy weighs almost three pounds, and two and a half of that is pure, bicep-crunching muscle. In fact it is temporarily the strongest muscle in the body, stronger than the heart, stronger even than the long voluntary muscles that run the length of the thigh and carry the whole body's weight in walking. There have been cases where a tightly squeezing uterus has broken the inadvisably inserted fingers of an obstetrician. That's strong. Encountering the uterus face-to-face the two-ounce cigarette-filter-sized cervix wouldn't stand a chance. The way the cervix in fact gets enlarged — so keeping itself undamaged for that crucial in-built, second labor ability — is much more subtle. The ultramuscular top of the uterus tugs on the bottom of the uterus, and that rather less muscular portion tugs in its turn on the cervix. But even that once-removed pull would be too strong if directed straight out sideways. So what's to do?

The impasse is that if the uterus pulls directly it is too strong to spread the cervix safely wide, while even if it pulls indirectly it's still too strong. What actually happens — and this sounds pretty bizarre but bear with us — is that the uterus actually uses its pulls to swallow the cervix. Each contraction, each resulting cramp and grimace for the woman, is taking place to pull the cervix up into the uterus itself. There's a reason for it. Imagine you're wearing a turtleneck sweater that is so tight it can't be pulled up over the head. One solution would be to grab a good hold of it where it's around your neck, and tug it heartily outward. That might make it loose enough, true, but it's also nearly certain to stretch it so out of shape that although it may still have some use as a floppy, open sweater, it won't be much good as a turtleneck again. That's about what would happen if the uterus just yanked straight out on the cervix, and as we've seen, it's no good because it would mess things up for a possible second birth.

The other way to get the sweater loose enough is just to steadily pull its neck up over the head, so that, bit by bit, it adjusts to the size of the head it has to let through. That's why the uterus pulls up the cervix. It's the only way of stretching it gently enough, gradually enough, so that it widens to the width of the baby-containing uterus and still doesn't get warped out of shape in the process.

The cervix, however, does not take this pulling without protest. When the uterus tugs, it tugs back. At first it doesn't matter much, not when it's the 2½ pound brawny uterus against the minute cervix. But once the cervix is a bit stretched, it's been tugged a little way up into the uterus it's like a spring being pulled wide: the more it's pulled, the more tension it has to pull back with. Even when its opening has enlarged to only one single inch, at a time less than a quarter way into the birth, it's pulling back on the uterus with force enough to crumple a small tin can.

So long as the uterus is actively pulling, that's not too bad a problem. But as is well known the uterus isn't able to keep up its tugging without a rest, and instead it contracts and rests, contracts and rests, all through the duration of the labor. Those resting points are when the gamely tugging cervix has the best chance of getting back, undoing in the rest interval all the stretching the uterus managed to do during the pull.

To ward off that springlike counterpull, when its guard is down, the uterus does something no other muscle in the body can do. Once it's finished pulling tight for a contraction and needs to take a breather, it lines up its muscle fibers like the two sides of a zipper, and then meshes them together as if the zipper were being zipped. Now it's safe. Now the cervix can gamely pull back as hard as it wants, but it won't budge the uterus out of position an inch. The uterus can take its rest in peace.

When the uterus is ready for another contraction all it has to do is unzip the perfectly matched tendrils of its muscle fibers. Then it can get loose with another straining tug, pulling along through the lower uterus and yanking the cervix exit just a little bit farther in. Zipped and unzipped the uterus muscles go, one full switch for each contraction of the labor.

In a first-time birth there are liable to be 150 of these contractions in all. Almost all will occur just to pull the cervix up into the uterus and safely to the side of the baby's head. After that only a few more are needed. In a second birth, the total can be as low as 75 contractions, in a third or fourth 50, and after that, 40 or under (though it is rare for the progression to be so simple). There have been cases of an entire labor taking place with the mother's feeling just two graceful contractions — one pulling up a much-used and eminently stretchable cervix, the other sending out what was only the latest of an already large family.

For some reason a number of American women are taught that many African women are able to have their babies with only a few contractions like this, perhaps squatting next to a furrow in a field and then can pick right up in the plowing where they left off, only this time carrying the newborn baby with them. It sounds impressive, but it's just not so. German anthropologists pushed this view in the early years of the Third Reich, when anything that showed how animalistic all but the Aryan races were was being gladly published in Berlin. Unlike most of the other claptrap that was invented to support the racist ideas of the Nazi party, this notion has somehow curiously lingered on. All that is really certain about birth time is that younger women tend to have fewer contractions, and quicker deliveries, than older ones, a rule of thumb being one hour longer for the main part of the labor for each 10 years that the mother is over age 25.

Old or young, the uterus's efforts to widen the cervix are helped by the baby in a most peculiar maneuver, which could be called Bernoulli bashing. It happens by using the delicate bubble of amniotic fluid around the baby as a power-driving battering ram. It sounds rash, but really it isn't: the name shows why. Way back in the 1700s a Swiss mathematician named Bernoulli (whose father, grandfather, uncle, brother and son were all mathematicians too) noted that however hard you push a fluid that's held in a container it's not going to get smaller, as say a marshmallow does when you push on it, but instead will just squeeze out through any available weak point in its container with the pent-up energy of all the pushing that's been going into it. This principle is the one behind all hydraulic control devices. (In a Boeing 747, or the Concorde, for example, the flaps at the back of the wing lift up after landing because the pilot has pushed a switch that squeezes down on fluid in a "tub" that starts just under his seat in the cockpit.) The same system is pushing away in the uterus. With each contraction the uterus slams down on the amniotic sac and its fetus inmate. That shove makes the fluid in the sac push out on the cervix just as hard as the uterus is tugging it in. Double power.

It shouldn't be thought that the fetus is just a passive rider in this Bernoulli bashing. If anything it's still the master controller of all the action. It was its hormones that sped out through the umbilical cord and started the contractions in the first place, and it's those hormones continuing to race out through the cord all this time that have kept the contractions going, whether the mother wanted them or not.

That uncontrollable nature of the contractions is behind the frequently discussed matter of pain. For male writers it's easy to sit back and be superior about how all the pain can be overcome with the proper mental attitude, but in fact a look at the strain receptors around the uterus and cervix shows that pain signals are going to be sent up to the brain whether one thinks them morally admirable or not.

From the cervix for example where the brunt of the stretching is going on, pain signals head out to the very bottom of the spinal cord, and then up along in there to the brain. Since the point where they reach the spinal cord is directly under where the mother is pushing, it's almost impossible to head them off at that switching junction. This is not good, because a stretching cervix hurts (it's usually the main cause of menstrual pain by the way), and in the few hours of labor it's getting the stretching of a lifetime.

Still there are other ways to cut down its pain. One is somehow to get the brain to send down the spinal cord outgoing messages that meet the pain signals halfway and zap them into oblivion before they can get up to the brain and register. Certainly the natural morphine endorphin chemical does squirt out to extra-high levels in the brain and spine of a pregnant woman shortly before labor begins. We'll see in chapter five how this works and it's the idea behind the many instructions a pregnant woman gets to relax, not worry about the pain, and generally try to keep it away by concentrating hard enough on how pleasant the whole experience really should be. There's a certain merit in this approach, for such states of mind really do send down defensive signals that block off the pain. But there is a limit to how many defensive nerve channels there are. A good number of the pain messages from the cervix are likely to get all the way up without meeting any blocking defenders at all. Also even the few defensive pain-blocking signals are not going to be sent down if the woman is anticipating or keenly apprehensive about the pain.

Veiled references about the agony of childbirth do nothing to make it easier. Such malicious comments seem to come most frequently from women who have few grounds of superiority over a younger one than that they have had a child and the other woman has not. Short women married to large men will have painfully large babies, they whisper just loudly enough to be heard. Invariably they're wrong. A short woman of course is in no danger of having a large baby because her husband is tall; the fetus's size has no choice but to match the mother's best capacity.

There is another way to block the awesome pain from the tugged and battered cervix. It turns out that before the pain receptors from it reach the spinal cord, they first twirl around in a series of tiny bunched circles almost like a miniature jumbled knot, that is situated above the very top of the thigh, just an inch or two below the surface. At this staging point a single shot of a local anesthetic can knock out the pain pathways with over 90 percent effectiveness.

The other source of pain is the contracting uterus and even trickier is trying to stop the pain from there. Again, it's different for each person, but there definitely are pain and stretch receptors on the uterus, they definitely do feed into the spinal cord and so into the brain, and they most definitely do fire when the top of the uterus is undergoing its mighty contractions. Some women feel them more or less than others, but it's plain false to say that the pain signals are really not there. They are, and the sensible question is what can be done to stop them. That at least is how it looks now. But until little more than a century ago, the question instead was merely how stoic the woman could be to survive the contractions without relief.

As an example just what this attitude could lead to, when the daughter of England's King George IV (then Prince Regent) went through an agonizing fifty-two-hour labor, from which she died shortly thereafter, none of the royal physicians in attendance did anything to help her, neither repositioning the baby, nor giving the mother a stiff drink, nothing. A painful labor was a woman's lot, and so the nineteen-year-old princess had to go it alone. (Some empathy did slip through, though. The chief physician who stood idly by shot himself in the head several years later during a similar labor, in torment at what he had seen.)

Relief of pain during labor became acceptable only when Queen Victoria chose to take several breaths of chloroform when she was delivering her son, Prince Leopold, in 1853. She had inherited the throne because of the death of George IV's daughter, and, since she had picked up intimations of the story since her youth, it was a most understandable choice on her part. Those breaths gradually made research into relief of labor pains acceptable. That research is what led to the understanding of how cervical pain is signaled, and it's also led to a helpful grasp of uterine pain too. Let mothers today bless old Victoria's fear.

It turns out that pain signals from the uterus lead into the spinal cord at a point in the lower back some four or five inches below the waist, a spot much more easily reached from outside than is the cervix's relay spot under the buttocks. That's where the procedure known as the epidural block comes in.

When the pain signals from the uterus pulse into the spinal cord, they spend a while traveling close to the surface and not at all submerged in its inner depths like many other signals. In this near surface skimming position they can be easily headed off by a hypodermic needle. A squirt of a mild anesthetic injected into the fat globules around them in their easily reached spinal cord position will knock these uterine pain pathways clean out of action. As an added benefit, the shot that does this will have no effect on the much more bundled up nerves controlling muscle contraction that pass through there, so the woman will be just as strong as before, on top of being pain-free. This is ideal for the lengthy first part of labor when the uterus is doing so much work.

Before epidurals were perfected in the late 1940s and 1950s a weird combination of morphine and other drugs was often given instead. It was rarely near as good. It frequently didn't stop the pain at all, but just wiped out the relevant memory traces in the brain. A few hours after the labor the woman would have no recollection of what had happened. It was called "twilight sleep," but as it merely made the woman groggy while the pain during the actual labor was as high as ever, it was really more of a "twilight nightmare." What's more if given too soon, the drug would stop the labor completely; if given too late, it would knock out the child's lungs and so suffocate him or her. Despite this it was easier to use than an epidural shot (which requires a skillful hand to hit the right place), and so was only gradually abandoned as the 1960s rolled on.

Still much of course can be done when the highly regulated setting for an epidural isn't available. One simple alternative is an anesthetic face-mask that dangles free next to the woman in labor; she can reach for it and take as many breaths as she wants whenever the pains are more than she desires. The mask supplies oxygen for the embryo as well as nitrous oxide (laughing gas) for the mother; if too much is inhaled the mother falls unconscious and the anesthetic mask slips away, so the nitrous oxide level in her blood goes down. Simple, neat, and effective. And only possible now that the guilt of avoiding pain is gone.

Finale

Once the cervix had been pulled up into the uterus and cranked open to a baby-allowing four inches wide, everything is ready for the child to come out. This is the quickest part of labor, lasting under two hours on average for a first birth, compared with twelve hours or more for the first part. Once again, the woman's body figures out a way to double the energy of the uterus. It can't continue to use the battering power of the water sac around the child, for by now that sac has popped and poured out. Instead, during this second part of labor, the woman is finally able to move the process along with conscious efforts on her own part. (All the contractions in the first part, by contrast, happened automatically, without her being able to stop even if she wished to.) Now she can squeeze down, hard, and when she does the muscles of her lower abdomen squeeze and squeeze the small space around the uterus, compressing it so hard that the uterus within squeezes too under the onslaught. To resist against it there's only a seven-or-so-pound baby, with a gaping and slippery cervix opening before it, and not even a slippery handhold it can grasp to keep from being squeezed out. So, inch by inch, out it goes.

All this pushing gets applied square onto the one part of the baby that is most deeply wedged in the uterus, the buttocks. What this leads to is clear from imagining what would happen if you were stricken with an impulse to lean over and press your head into the soft cushion of a couch, upon which moment someone started pushing on your now proudly raised posterior from behind. The stress would go straight along your spine, and jam your head even harder into the couch until it finally twisted down towards your chest, and probably a little bit sideways if the push wasn't exactly straight. The baby is in the same predicament, and suffers the same result. It's getting pushed from behind, but its head is wedged against the tight, final inner rings of the cervix. So down and a bit sideways goes its head too.

That's a most gratifying change. It shows that the end is near. For now the baby can slip straight into the downward-tilting vagina. As it starts to go there's that much less of it jamming against the ultra-painful cervix.

Down along the tunnel of the vagina the baby goes, keeping its head neatly tucked down and neck slightly twisted, until the very last moment. Then, barely an inch from being out, the head gets twisted and cranked once more. It happens because the back of the head catches tight on the uneven protrusions from the lurking bones of the

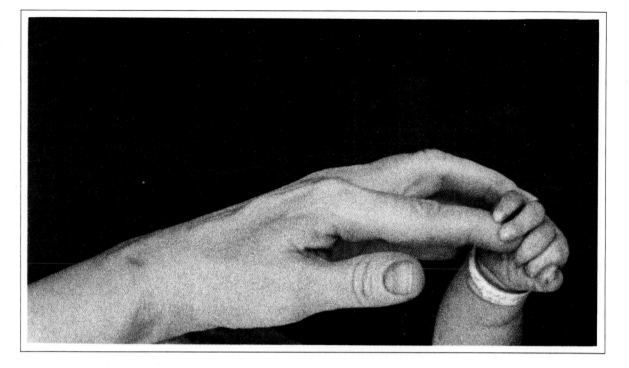

The loving touch of a now inde-pendent life.

pelvis. That levers the head back up, which is fine, but unfortunately the head is cranked sideways even more.

With this final shift, just a moment before birth, the baby is in a most extraordinary position: its feet are still up in the uterus, its torso is twisted around the right-angle bend leading into the vagina, and its head is cranked sideways to an amazing extent. Jumbled around in this gymnastic pose the baby presses with the next uterine contraction right into the sensitive inner wall of the vagina, the perineum, and not out along the exit path where it had moved with all the other contractions. That produces what is often remembered as the very worst pain of the entire pregnancy, a sensation described as feeling "as if everything was splitting wide open." Nothing of the sort actually happens of course — remember how the setup's rigged to stay together for at least a second child — and this most intense pain is brief.

Only two or three more contractions are needed to wriggle the baby out the remaining half-inch after that. At one of those pushes, no mightier than the others, the baby's head finally pops out. It's so sudden that the mother is almost certain not to know of it unless someone, most often now the near-hysterical father, tells her. The baby by contrast knows very well when its head is out. Without a second's hesitation it firmly uncranks its twice-twisted neck. That's a meaningful sign. Now free in the world, it's going to be boss.

Painfree and deadly: image with short-lived radioactive particles of a massive brain tumor which had grown totally unsuspected to the patient — for the brain has no sensors for detecting pain within itself at all. Main image is side profile; upper one positions tumor on side.

5
Pain and Illness

Pain

At one time or another most children believe that by closing their eyes tight and wishing hard enough they can cause anything to happen. In this curious belief they bear a striking resemblance to most adults, who also at one time or another, know — really know — that their favorite beliefs must be true, and are disinclined to let the imposition of disagreeable facts spoil this delicious satisfaction.

For ordinary adults, as for ordinary children, such moods do not last long. One ideal after another that it would be desirable to hold true is undercut by the way things really are. Bosses do not always recognize initiative, lovers do not always recognize affection, and perhaps most personally of all, the body we inhabit does not always act as effortlessly or painlessly as it could.

Pain is a nuisance, a burden, an agony, an affliction. Nobody likes it, though everyone gets it, and what good it does is often beyond comprehension.

Some say pain is granted us so that we can avoid accidental damage — if that's so, it's a remarkably unwieldy grant. The pain produced by an accidentally bitten tongue, or a suddenly stubbed toe, can be agonizing and startling out of all proportion to its significance as a marker of injury, while the truly ominous growth of a breast cancer or brain clot can be silent and painless until too late. Also, most of the reflexes protecting the body against injury work perfectly well without producing a sensation of pain, such as blinking, running, or the adrenaline surge during a fright.

To find out just what this peculiar experience called pain is, to get at the sheer feeling of it, we'll need to begin with its underlying mechanics. These have been at least dimly understood for a long time, and were given a key boost by a tribe of Germanic warriors of the Dark Ages usually more remembered for their military deeds than their science: the ancient Franks.

Mechanics of Pain

The contribution they made was unintentional. When most Frankish warriors got mad at one of their fellows, they were likely just to hack at him with their swords. But for the few Franks who believed in outlining their acts with linguistic exactitude, the hacking was likely to be prefaced by the remark that they were going to strike the offender, a bellowed remark that sounded something like "hurten." Picked up by the frightened peasants in the region, and transmitted from them into French and later into English, it's the origin of our word *hurt*. The harder the Franks struck, the more they would have agreed that they were hurtening.

And that's how scientists agree most pain receptors work. The harder they're struck, the more they hurt. There are, of course, exceptions. These pain receptors are scattered throughout the body, with the ones responding to the lightest touch, and so producing the least "hurt," being placed in the topmost layer of the skin. If you look at your fingertips you can probably just make out the ridges that show in your fingerprints. In the valleys of each one of these there are two or three special little branches sticking straight up; others are weird bulbous capsules jammed full of nerve-connected receptors. All are quivering in readiness for something to push down just a bit too hard on the skin overhead. If that push comes, the quivering ends are squeezed down too, opening numerous fissures in the wall for electrified sodium atoms to come sluicing into the little hollows inside them. Those quivering ends are connected to a long tendrillike pain nerve fiber, and *it* doesn't rest unchanged when the top gets split open.

The leak at the end affects potassium-laced water farther along the pain nerve too, and that further splashing, as we saw in chapter two, sends disturbances along the channel that forms the center of the nerve fiber. By the time the signal reaches the main body of the nerve, deep in the center of the finger, it's traveling at over 80 miles per hour. To reach that extraordinary speed, the nerve fiber has fat sticks on its outside that the signal can use to skid and leapfrog from one spot on the fiber to another, so gaining speed with each jump.

This signal would never bother us at all if it just swirled around in the center of the finger. Instead the pain signal follows the thoughtfully arranged nerve fiber — which stretches on, up the arm and around the armpit (a region which, as all nerves from the arm shunt through it, is surprisingly sensitive), and on to the spinal cord. Bubbling away in the extreme end of the nerve here are eager granules, Ping-Pong-ball-shaped but ever so much smaller, that are filled with the arch-evil amino-acid droplet known only as "substance P." When popped loose, the tiny balls let the ominous substance P out loose into pickup nerves stretching vertically up the spinal cord into the brain.

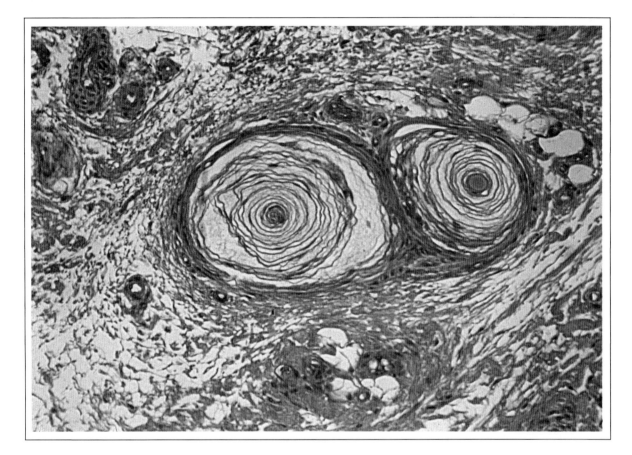

Two deep pressure receptors in the skin — as wide across as a single fingerprint ridge. Free nerve ending at center fires when concentric rings are sufficiently compressed, as in a deep paper cut or crushing bruise. Note fat cells at upper right edge of the smaller one.

Now they fire, and that finally is what hits the brain with the irritating message that there's pain. Ouch!

That's only step one. If the original source of the pain signal continues to push even harder and deeper into the finger, then, as the Franks recognized as they grunted out "hurten," it will hurt more. The reason is that under the first extended, quivering receptors, about five times as deep inside the skin, there is a whole other series of pain-sensitive nerve endings. This second set looks like a bunch of tiny cotton fluffs, and, when these are pushed hard enough, say by a neatly slicing paper cut that makes it into this middle layer of the skin, they will send off a series of signals through the nerve channels that run next to the ones of the surface receptors.

Because of their different beginning, signals from the types of receptors are interpreted differently when they reach the brain. The first kind produces a tingling or itching type of pain, while the second, deeper receptors go on to produce just that sharper, burning pain that an inadvertently slicing envelope can so unpleasantly produce.

Even that's not the only kind of pain the far from stoic skin can create. Most of us have experienced the heavy, throbbing feeling that comes from a deep cut or an unblocked blow, and that type is produced by yet another set of receptors. This third grouping is wedged down under your skin even lower than the others, way in the squishy

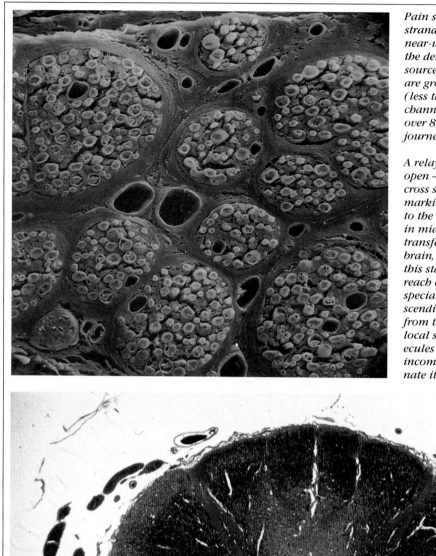

Pain signals head on. All these strands carry chemical surges in near-unison when pain strikes the detection units at their source elsewhere in body. They are grouped in nested bundles (less than 30 separate nerve channels in center right one, over 80 in upper left) during journey to spinal cord.

A relay station that's not always open — the spinal cord seen in cross section. Incoming pain marking bundles (in red) reach to the butterfly-shaped portion in middle. Instead of being transferred straight up to the brain, they can be blocked at this staging point, and so never reach consciousness centers, by special antipain impulses descending down the spinal cord from the brain which causes a local supply of enkephalin molecules to fit onto the end of each incoming pain nerve and eliminate its signal right there.

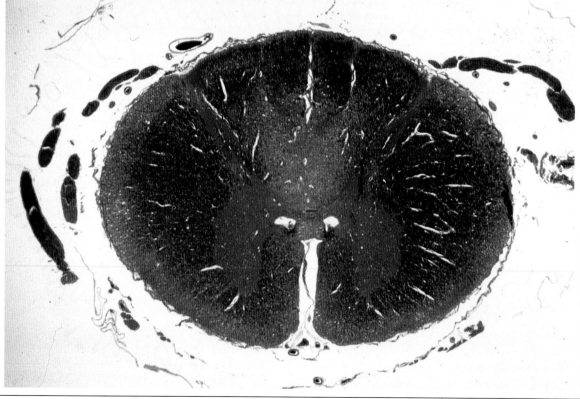

yellow globules of fat that form a cushioning and insulating layer between the inside of the body and the skin.

These innermost skin pain receptors are giants in the subcutaneous pantheon, stretching sometimes as wide across as a full fingerprint width, and always being covered with layer upon layer of thin spherical shells. Protected with that armor the one free nerve ending at the center is not going to be triggered unless something really drastic is happening in the vicinity, and even then it takes a while to start up. But once it does, it fires with all its worth. This impulse, just like the other two kinds, races up the arm and around the armpit before reaching the spinal cord. Once it gets there it does not have to reach in far before special using circuits back there in the spinal cord take its message carried by stocks of the old nasty substance P again, and shunt it on up to the brain. Few bangs or bumps are strong enough to trigger this final sensor, but for the ones that are, such as a car door pinching a finger, the pain signal that gets sent off this third way is strong.

There are other pain sensors like the three finger types scattered throughout the body, and they usually produce one of these three kinds of signals so automatically, it would seem that the same bumping or pricking would always produce the same amount of pain. But that of course is not how things work: the same event does not always produce the same pain. It's because the body has defenses.

A muscle-bound hulk of a boxer may be totally unfazed by tremendous wallops to the jaw from another muscle-bulging hulk when he's in the middle of a match and his pride is on the line, but he's quite likely to have shrunk back in agony and terror a few days before when a stoop-shouldered, white-coated dentist prodded gently around his gums in an antiseptic office, with the whine of a drill in the background. All of us know how distraction or force of will can keep a pain from being noticed, as when you ignore hot coffee spilled on the hand during a job interview, or when you notice only at the end of a movie that you've got an ache in your neck from looking up at the screen. If the nerves that send out the pain signals work as automatically as described this should never happen.

The explanation is that the pain signals may always get started on their way, but there's no guaranteeing that they'll always get through to the parts of the brain that do the noticing. The body defends against them with several fallback points of resistance. Go back to the signal from the hand that's just rounded the armpit and its coming into the spinal cord. The slender channel along which the 80-mile-an-hour wave is traveling has stretched without a break all the way from the finger. It's in that backbone-surrounded column that the first interference with its arriving can come.

Increased activity in the right parts of the brain will send swarms of electric signals down into the spinal cord, some of them flashing head-on into the pain signal coming up. Over one million nerves are always on call, leading down *into* the spinal cord from the brain, precisely for the purpose. When they meet the incomers there is likely to be a spark release of stored molecules, called enkephalin, which fit onto the end of incoming pain nerves and smother their signal, so stopping any pain impulse from getting on up to the brain. No input signal, no sensation of pain — and so no problem with continuing the prizefight, going on with the recital of the job candidate's qualifications, or neck-loosening contortions in the middle of the film.

This is how we develop the pain maps that divide up our bodies into sensitive areas where pain has already struck, and not-so-sensitive areas, which don't have any connections with prior pain. It's why someone who has had shoulder surgery years before which has totally healed is likely to flinch when a lover's caress reaches the top of the arm, while someone else who might barely notice a hefty bump on the shoulder is likely to act quite irrationally before a slight stomach bug if he had suffered unpleasant bouts of nausea long before.

These pain maps really exist. There are extra sensing channels in areas that were overtriggered once before, and many of them end in receiving tops that are enlarged and especially easy to trigger.

Pain maps don't even have to come from direct experience, as middle-aged men demonstrate when they mistake a sudden stomach pain for the dreaded angina they've been warned about. (The stomach is actually much higher than most people realize, dangling almost way up in the chest not far from the heart, rather than wedged down low in the belly.) No signals are going to be sent down from the brain to ward off these pain signals when they come up, for that's what it means to be keyed up for a definite pain. The result is that it reaches the highest awareness centers in the brain at full speed and with full impact.

Put an electricity registering device on the back of the neck of an average prerson who ate a bit too much, and if it were sufficiently sensitive it would able to play out a geiger-counter–like storm of clicks. The rushing of the usual blocking-off antipain nerves coming down from the brain. The same device on someone who's constantly fantasized about pain from the middle chest will ring out with no defensive clicks at all. Without this defensive barrier the path is clear for the actually innocuous minor pain signals to rush up with full intensity.

Now for fallback defense number two. This is when not enough countering impulses get down before the pain impulse gets up. Even then it's not too late to avoid some of the pain. A netlike structure at the base of the brain which channels all incoming signals can squelch out morphine imitating molecules that halt the pain impulses from crossing this final gate, and keep them from entering consciousness.

How to bring on this second defense? The excitement or rage of combat can do it, as soldiers who withstand great wounds in battle without problem are probably showing. Their reaction is the same as someone doped with morphine — fitting because the brain's natural molecule reserves here actually have the same 3-D shape as morphine, and so affect the same neuron leads in the same way. Even in ordinary life concentrating hard enough on something else can do it — the labor section of chapter four, "Conception and Pregnancy," discussed it in that particular case. That's why it's useful to whistle when you stub your toe, or why someone trying not to feel dental pain can be well advised to clench and unclench his fingers. Concentration on such peculiar tasks is likely to help keep the brain's usual circuitry from being ready to give the full treatment to the painful impulses — as well as bring out the natural morphines. This demands some fortitude though, for the slightest pause in the new concentration will let the eagerly inward-swarming pain signals arrive in all their unwanted intensity. Even the cerebrospinal fluid sloshing around in the brain, that we met in chapter one, can get poured full of these personal morphines, and used as a backup channel to bring them, waterfall-like, down to the rest of the body.

There's a final, third way incoming pain is defended against. It's one far more easy to apply. All you have to do for it to happen, in fact, is to sit back and wait. This works because a lot of pain that is usually felt comes not from the special pain receptors, but from the normal sense receptors that are scattered through the body alongside them. And they fade out before a constant stimulus. An example will show how.

Back in the top layer of the skin, next to the cotton ball-like nerve endings that detect pain, there are also little ball-like nerve endings that detect cold, and fernlike ones that detect heat. Normally these ordinary receptors produce readings of hot or cold; mild sensations nowhere near as overwhelming as the signals sent in by their neighbors, the pain specialists. If that's all they could do there would be no way to get the sensation of being too cold, or being burned. Since they can't change the quality of their signals to be automatically felt as pain, like the usual pain receptors do, all they can do is send in more signals than usual, and send them a lot faster than usual. That's the source of the burning feeling of local freezing pain: just the usual temperature receptors sending their impulses on to the brain in quicker succession than usual.

This kind of pain will go away the moment the firings become less frequent. One way is to wait until the burn has totally cooled down, or the chilled fingertips have totally warmed up. But that can take quite a while, and it turns out that it isn't even needed. The nerve endings that send in these temperature signals begin to slow down of their own accord once they're presented with an unchanging source of heat or cold. This happens because all the nerves in the body, except those exclusively concerned with pain, fire at their utmost only when they're triggered by a changing sensation. Grab a pencil in your hand and you'll easily feel all its contours on your palm. Continue holding it in the same position for half an hour, and if you stop to answer the phone, you might find yourself looking all over the place for the pencil before realizing that it's still in the steady grip of your hand. The nerve endings that registered its being there stopped firing once the impulse was no longer changing. That's why all the pain sensations that come from normal channels being overstimulated quickly fade away.

The Sensation of Pain

So much for how the signals that we perceive as pain work. But why do we perceive them as pain; why is it that they strike us with the special sensations that makes us say they "hurt"? Couldn't they somehow travel the same pathways from the scattered receptors yet make us feel good instead?

No. If they felt good we would be able to ignore them. The fact that they're painful ensures that we don't. It would be hard ot believe someone who said, if you asked him how he was, "Oh, I'm fine today, but wait, come to think of it, I do seem to be suffering agonies from a body-wracking pain shooting through my chest." The successive barriers of pain defenders do a lot, but they never allow such insouciance about even a much more minor pain. Because of pain's forcing itself unpleasantly on our attention, we are forced to be aware of where our bodies are, and how they're being acted on by the outside world. In other words, we're forced to take account of reality.

That's why pain can't be remembered perfectly, either. You can remember having been hurt by a pain, but you can't close your eyes and bring back the sensation in its full vividness. Orgasm is like this, too, for if people could remember an orgasm as vividly as the orignal sensation of it, they would have little incentive to seek to duplicate it. Similarly with pain: if by memory it could be brought back as fully as when it really happened, then the real sensation would lose its overwhelming power.

Not all creatures have the pain signals to ensure that they stay in touch with so many portions of reality. A contrast with some invertebrates shows this. They just don't have enough brain space to register pain, and so they can't fully hook up with what's going on in their bodies or in the world in which they live. In certain worms, cutting off one end will leave the rest of the body wriggling merrily along.

Moving way up the evolutionary scale to the likes of the mighty ostrich, things here clearly become different. Although the ostrich is not without its own peculiarities, having something like a heart down in its pelvis to pump lymph, it does have a big enough brain, and enough of the right circuitry leading up to it, so that if *it* bangs against a rock, knowledge of this goes up the brain and becomes a definite sensation of pain. The result may be a carefully considered body-straight, knees-high sprint into the desert away from the offending rock. It suffers pain, and so it's aware of itself and the world. Humans fit in the ostrich category in this respect.

The only way to be fully conscious is to have pain receptors from the whole body insist on getting the brain to register their messages, even if the brain would be happy not to know about it. In creatures that have it, such as us, this linking is hard to break. When it goes, personality often does, too. Soldiers wounded by the very nasty AK-47 in Vietnam, if given enough morphine by the squad medic, would often report that they "knew" they had a pain, but weren't worried about it and didn't really "feel" it. It's no surprise that when they were highly doped they usually also had only the vaguest idea of their personal identities.

More confirmation comes from the history of lobotomies, the practice of removing the front part of the brain, to end severe depression or rage. Sometimes they worked, but usually they didn't. Lobotomies became a minor fashion in the U.S. in the 1940s and 1950s, some

50,000 being performed. Techniques included jabbing an icepick through the inner corner of the eye, pushing it into the brain, and then, in the words of one horrified onlooker, for several minutes having the inserted icepick "wiggled, tapped, wiggled." It destroys some key links between pain signal inputs and consciousness.

Those who had this operation reported that afterward they were aware when something painful happened to them, but that it didn't bother them, and certainly didn't hurt. They too often also had only the vaguest sense of their own personalities, and the existence of their bodies and the world.

That's how much one is at a loss without the forced feedback from the sensors of pain. Some idea of the loss can be imagined by reversing the situation and remembering the common experience after being away from a mirror for a long time, say after a camping trip, of being startled at how peculiar your own body looks when you first see it in the mirror again. For somebody incapable of any pain, that startle of recognition would never come.

The special quality of pain divides out another way, too. This is the difference between pain that you feel is happening to some restricted part of you, and pain that somehow seems to be an integral part of you. A sore wrist or even a broken leg in a cast are good examples of the first kind, for someone with these ailments is likely to say he or she feels fine "apart from the leg" or wrist. It's somehow an impersonal pain.

But severe menstrual cramps or a very queasy stomach is different. This kind of pain is no longer impersonal, and there's no "rest of the body" that can feel fine while it's going on. A landlubber in a tossing boat is likely to feel bad, period, just as someone who has eaten a rotten oyster feels bad all over, too.

The difference between impersonal and personal pain is that the former is somehow not hitting at the center of our most general pain maps, while the latter is. The pervasive feeling of personal pain has nothing to do with its location, for a small cut on the chest, right at the center of the body, is likely to be felt as just an external, impersonal pain. Nor is the mere spread of the pain enough to make it into a "personal" one, for someone with a wide-ranging skin rash can still feel that *he's* okay, and that it's only his skin that's a mite itchy. The real difference lies in the different ways pain and other sense receptors are distributed throughout the body.

One group, such as vision receptors in the eye or heat receptors on the hand, is always near the surface, and spends its time feeding in signals from the external world. Any overexciting, accordingly, is likely to be thought of as somehow external, and not a "real" ailment. The other main group of receptors is distributed all about *inside* the body, and it measures how the body itself is doing, things like stretching of the stomach, or inflammation in the muscles. There are hordes of these inner receptors resting everywhere like long-forgotten bits of wiry fluff, under your ribs, behind your nose, by your elbows — and with a total of many millions everywhere else.

Most of the time they check what's going on without making any

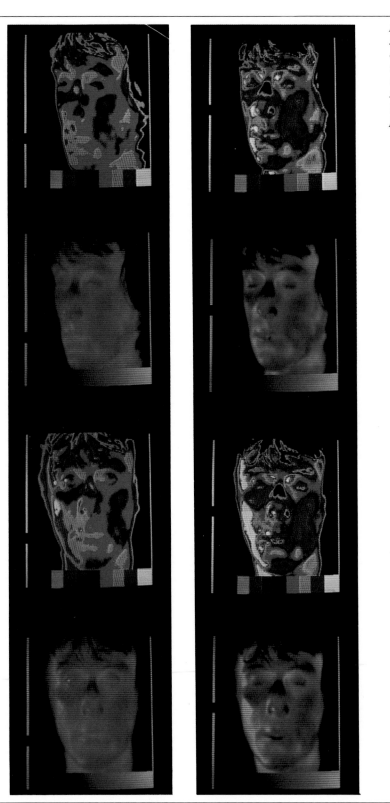

Face thermogram. The constantly changing heat patterns on the face indicate minute muscle contractions and shiftings, which through feedback sensors help give us the sensation of living within — encompassed by — a body.

fuss. Their tack is so boring in fact that usually their readings are shunted straight to the bottom or back of the brain, never passing through any consciousness centers at all. Who, after all, cares about such readings as "nose is straight, nose is straight, nose is straight," or "fourth thoracic rib fine, fourth thoracic rib fine, fourth thoracic rib fine"? These receptors take such a comprehensive and dumb interest in your body's personal property that they have been given the name *proprioceptors,* from the Latin for "receiving from one's own."

Very often, though, it does pay to take notice of them. Suppose you want to get out of a chair. As you move you can tell how much your ankles are straightening even without looking down, just from the constantly monitored level of signals shooting into your brain from the tiny stretched tendrils of the inner receptors winding all around the muscles in those joints. Without them you would have to check your ankles every time you wanted to know how you had to move to stand up.

At all joints in the body where there's a lot of turning movement, be it shoulder, finger, or knee, the bones at the center of the turning are kept from directly touching each other by a slick covering of a slippery fluid. Spread between what would otherwise be two regularly grinding joints it is indispensable in keeping down the body's wear and tear. A bit of this fluid sprinkled onto the bones would just dribble off to the sides and out of place. To keep it where it's needed, all your important joints have finely molded sacs fitting around the joints, a little like tautly stretched plastic bags.

That's where the trouble comes in. Illness, bumps, even general tiredness can make that lubricating fluid swell up. Most of the time it's not enough to see, and even a good doctor might have trouble measuring it. But the special sensors spread all along the fluids restraining bags have no trouble picking up the increase at all. They're in the right place, and of a small enough size, to send the word of this swelling up to the brain with the utmost clarity. Would that they didn't for when that happens, you ache, and the feeling seems to come broadly from deep inside. (Some of those receptors respond directly to various internal chemicals the body releases after an irritation or lesion. The way aspirin works quite simply, is to keep one of those irritant chemicals from even being built up at all.)

Many of the other deep receptors produce a corresponding deep pain, and they're at the very source of our feeling of having a body with an inside, and not just a loosely flapping surface through which we can measure heat or sound waves outside. Strong signals from these inner sensors are what produce the feeling of something going wrong with one's very essence, something undermining not just the working of a part of the body, but the health of the body that goes to make up a person as a whole.

When that happens we feel most definitely ill, and the nature of that state, and the various things that can go to cause it, are worth a lengthy look.

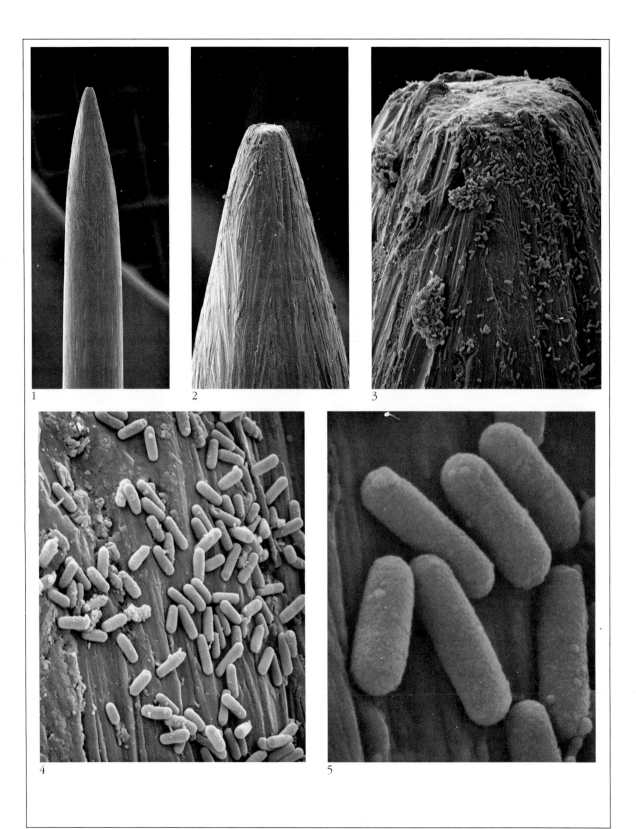

1

2

3

4

5

Omnipresence of germs — successive closeups of the point of a carefully cleaned pin. Fig. 1 is the tip of a recently sterilized pin; fig. 3, 25 times closer, shows immense numbers of rod-shaped bacteria that find their way there from the air. Note how some form clumps, while others stick impartially in line with minute grooves on pin or across them. The rubbing of some bacteria on each other in figs. 4 and 5 (continued greater magnifications) is not arbitrary: they are capable of transferring their genetic material and increasing their numbers that way, as seen in fig. 6. Equivalent closeups of a fork, spoon, or the rim of cleaned glass would show similar bacteria populations.

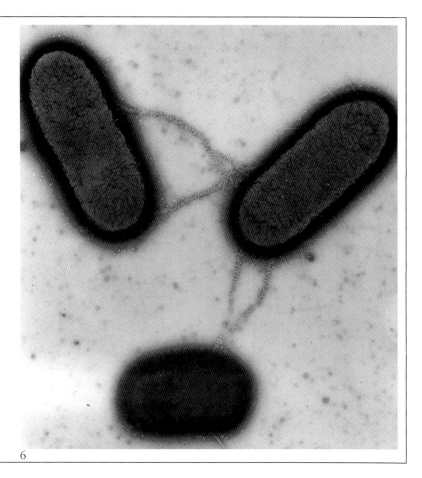

6

Illness

Being ill is not as simple as it seems. First you have to get there, and the entrance requirements are nowhere made clear. A health-food fanatic in Idaho would thing something was terribly wrong with him if he woke up coughing in the morning, but a chain-smoking writer in New York would be used to regular coughing in the morning, and would be worried only if it developed into the choking spasms of bronchitis. For an Appalachian coal miner, even a bronchial cough might be nothing to get excited about, since he's likely to have had one and to have gone on working in the mine for twenty years. It's relative.

These two sides are well seen in the origins of two common words, both of which developed in the Middle Ages in France. Burgundian noblemen then knew there was a special sensation to feeling good, and they called that sensation *aise*. (It persists in our word *ease*, as in "take your ease.") But they also knew that there was a clear sensation to not feeling like that and they called it not *aise*, but *desaise*. It's only a slight spelling change to our own *disease*, and the key idea in it has remained the same. You feel disease when you feel "not good." That's the subjective side: how it looks to the sufferer.

The other side comes from the fact that this feeling will be different depending on what "good" was to start with. That's why with slight flu you're never quite sure if you're ill or not. At work, where everyone's treated as if his or her health never falters, even fairly strong flu is likely to be ignored but back at home, where you're used to examining any sensation that's different from the firmest good health, flu is likely to send you into a pitiful pajama-clad, chicken-soup-swilling state. In the perspective of someone other than the sufferer, such as his family that has to put up with it, the decision to be ill can look highly arbitrary. They're taking a colder outside view, and that has an unvoiced side well expressed by the origin of the world *invalid.* The term has passed through the Medieval French and come from the Latin *invalidus.* The second half of it, *validus,* is related to "valor." But the prefix switches the meaning, making *invalidus* mean "not valorous." Any long-suffering household member who thinks that the flu sufferer being so assiduously nursed could cure himself if he just showed some more valor and pulled himself together, is very much holding true to the derisive origin of the word *invalid.*

Infection

How do we arrive at this sad state? The bugs that cause the slight infection of a light flu or cold, so placing us in the moral dilemma of whether or not to be ill, go about their job in a most distressing way. All of them are parasites that would put even the most cold-blooded gigolo to shame. Given the right supply of food they will grow and grow until they've used up all the food and are choking from each other's weight. Let them into the body through a single sniff of the air that they're floating around in or by a barely noticeable scratch, and they will most gladly try to do the same. And by the same portals cold or sore throat germs enter, can also come the less delectable invaders behind anthrax and pneumonia, tuberculosis and bubonic plague.

They are all the worst of guests. Microbes have an absolute fetish about having sex when they've landed in their new habitat. Not just once or twice, when their hosts are distracted elsewhere but hundreds and hundreds of times every twenty-four hours. The progeny of these acts are similarly inclined, which is why a cold can seem to come on you in its full-fledged misery just a few hours after your first noticing a slight soreness in the throat. More potent germs are no less quick; bubonic plague would strike dead its medieval victims in extraordinary agony sometimes just two hours after they had been feeling perfectly healthy (which is why the ring-around-the-rosy children's game ends with such a sudden fall: it's an oral tradition continuous from medieval times and the "rosy" plague sufferers).

These unwanted visitors come in two types. The first are visible only in a strong electron microscope, and will attack many living forms in addition to man. They're tiny but swift and in some life forms look like ultraminiaturized lunar landing modules, with long spindly legs topped only by a cannisterlike chest. These are the viruses — a word that many Roman empresses may have muttered to themselves as they scurried around in the food-storage room at night, for that was

Team of viruses on the attack. Above, having landed on a bacterium, and injected their genetic material through its outer wall; below, bacteria splitting open as freshly replicated viruses which have grown inside it from the injected genetic material burst out. Analogous microbial warfare will be taking place inside of you at heightened level every time sections of skin are rent open by a burn.

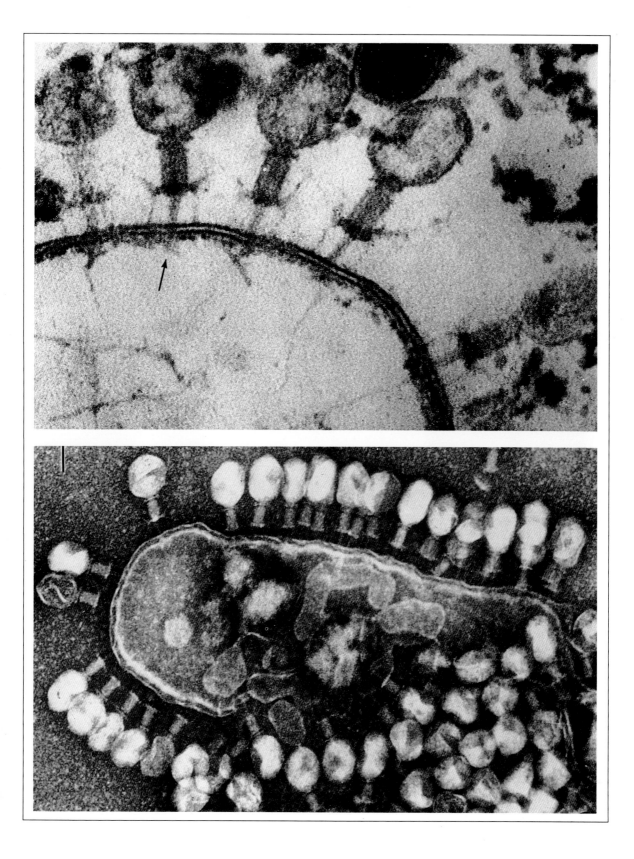

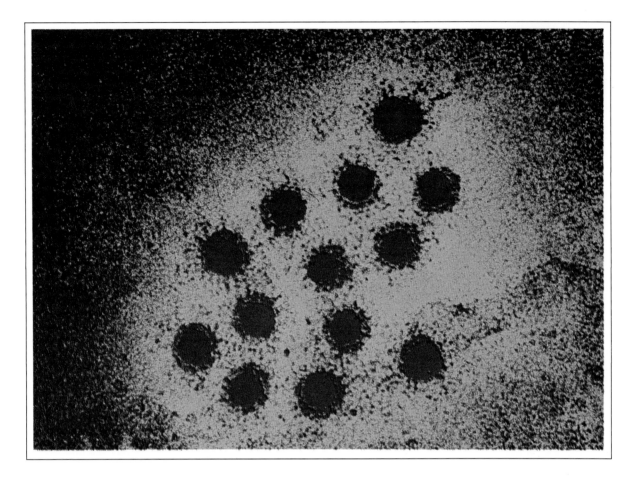

the way to say "poison" in Latin. A description concentrating on bacterial viruses will be most illuminating.here. Tumbling about in the
bloodstream, the viruses only go into action when they bump into a
healthy cell, an occurrence which, considering that there are something like 200 billion cells in even a scrawny body, is usually less long
in coming. Smacking into a cell, this particular intruder will rotate till
its "legs" are pointing down, then plop them onto the surface of the
cell. Once the legs are in place, the sharply edged "chest" starts coming down, as if pulled by a whirring electric motor. Down to the surface of the cell it evenly sinks, splatting as it contacts.

Now the virus is in position to do its dirty stuff. With a squeeze of
its Meccano chest everything that was in it goes pouring out into the
wholly innocent targeted cell. A rush of livers, kidneys and coiled intestine is spared the cell, for the virus intruder squatting intemperately
on its outer wall is not equipped with such amenities of a proper life
form. Not even a shapeless stomach like the inside of an oyster comes
squirting out. No, all the virus has to show when it squeezes its chest
space hard is a single long strand of DNA molecule. That's all it had inside its bare 3400 ångstroms of length. This DNA is the much publicized master molecule of life, the same in all living forms, from which
an exact copy of its owner can be cloned.

*Serum hepatitis magnified
70,000 times. This easily transmitted virus, present in nearly
all countries, attacks the liver
with sometimes fatal effects.*

Shot on into the cell that the virus chanced upon, this DNA molecule goes quite determinedly about its business. Sometimes a virus won't even bother to pierce through the wall, but will enter the cell by just surreptitiously stowing away in the bubbly food transport chambers the cell always has ferrying in from its surface! Ten minutes after entering, hundreds of new empty chests and legs get formed, each one just like those of the original virus. Often an infecting virus will neatly stack up its future chests and legs on ledges inside the membrane of the target cell, where they can be slipped into at ease. Another ten minutes later the empty chests begin to fill up with fresh virus DNA that also is being formed. After thirty minutes it's done: hundreds of neatly cloned copies of the original virus come streaming out of the cell its empty casing is still perching on.

For the original virus that was great, but for the target cell the labor of making those clones is enough to drain it of all energy it had. Since all its juices and supplies go into making those copies, the cell has nothing left inside to live on. It pops open amid its unsuspecting neighbors, empty, wasted, and dead. The legacy it offers them is a nasty one. Those hundreds of exact copies of the original virus intruder, flying out of the shattered cell all shiny and clean, overall ready like their parent to feast on fresh cells. This is what's been going on inside you when you feel colds or flu, mumps, measles, or herpes.

The other type of unwanted microbe visitors are bigger, less streamlined, and a great deal uglier. These are the bacteria, some pliable, and near-octopus shape, some furry with millions of stubby waving hairs. They multiply incredibly quickly: many bacteria will reproduce up to 1500 percent in a single one-hour period inside you; if placed on a tray with the right nutrients a bacterium doubling every 20 minutes would weigh many times the earth's weight after only two weeks were up. It turns out that the speed of this reproduction has an incredible evolutionary tie-in with the entire origin of life. Here's how:

The DNA molecule in every human cell is not just a precisely coded rendering of everything that has to go on in the body. No, our DNA also has all sorts of "garbage" genes strung along as part of it — whole sections of the DNA strand that aren't any good at controlling any conceivable cell activity, but which just sit there; useless, filling space, and slowing down the efficient recopying of the DNA whenever one of our cells needs to grow. That's what it seems to be. But in fact it has turned out that those apparently useless, fouled up sections of our DNA encyclopedia-strands actually do have an important role to play. What they do is allow for more alternatives to be made when the DNA gets together and reproduces. Not just a few: they provide for one million up to one hundred million more alternatives than would otherwise be possible.

Billions of years ago, in the early Pre-Cambrian period when the earth was still warm from its original creation, all cells of the then primitive life forms had such junk in their DNA. That gave them the important asset of being able to evolve quickly and with all these alternative permutations, though of course because of the complications of reproducing when all that junk was around, it did slow down their re-

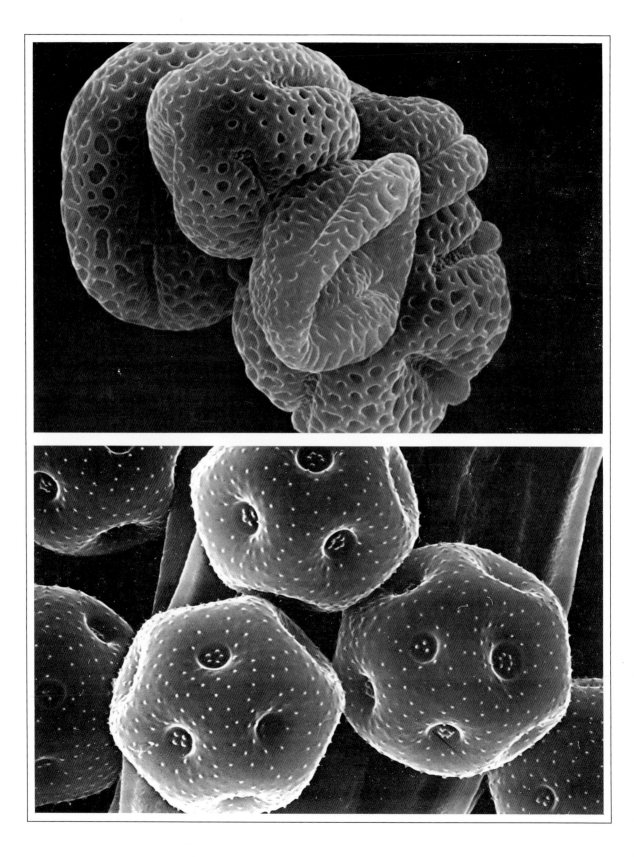

Pollen. Above, from lemon tree; below, from carnation. The bane of hay-fever sufferers, and a prime cause of disease-spreading sneezing. Rounded shape is to maximize internal volume; gnarled surface increases water containment — yet both these design features irritating to sensitive tissue in nose, producing inflammation, or 50 mph+ sneeze, or both.

production rate a bit. Almost all cells kept that setup — they were the ones that went on to produce the higher plants, animals, and man.

But some cells, ah some very special cells, dropped all the nonsense segments in their controlling DNA strands. This meant that they could reproduce faster than all the others, but it also meant that they could not mutate and evolve anywhere near as fast. Those hasty, cleaned up cells just stayed as they were, reproducing in near identical form, in one quick blind generation after another, all the way to today. And what do we call their descendants? They are nothing less than the bacteria that invade inside us today: living remnants of that tiny evolutionary switch billions of years ago. They are our long-lost cousins, showing what we would have been . . . if our ancestral cells had made that tempting DNA-cleaning switch. A close shave indeed.

There are other ancient switches that the evidence of bacteria point out to us too. Inside almost all your body cells there are submarine-shaped small objects that float around. They're in your arm and hands and fingers and even toes. Those streamlined floating objects help provide the power source for whatever the cell needs to do.

However did they get there? A distant choice is again involved, for it seems that they are descendants of an ancient, ancient bacterialike creature that got swept into an ordinary cell, and instead of getting dissolved, deftly managed to stay there and live. For in something like an invasion followed by a truce, it quickly entered into a mutually useful relation with the cell: it gave the energy, the cell gave it a home. Those submarine-shaped bits are a living survivor of that ancient invasion and truce: alive, and well, and inside almost all your body's cells even today.

Despite these deeply linked connections, present-day bacteria invaders have the bad manners to never stop leaving behind a dribble of unpleasant waste products called toxins. The name comes from the Greek *toxon,* meaning bow, and refers to the charming habit the early Greeks had of using their bows to shoot poison-tipped arrows. Achilles in the Trojan War was a notable victim.

The dribbling out toxins from bacteria are stronger than even these poisonous arrow darts. The toxin squirted out by the tetanus bacteria is as effective as strychnine, even in a dose a full 100,000 times less. Other bacterial dribbles are what can cause food poisoning, blood poisoning and many other diseases.

The germs that can do this are around all the time in absolutely staggering numbers. Each day we all flake off about ten billion little bits of skin, and on about two thirds of those quietly soaring rafts are living bacteria that could infect friend or foe alike. Even a thoroughly gowned surgeon has a million living bacteria floating loose from his skin in an hour — bugs which burrowed deep into the sweat glands, or waited deep in the curving caverns of the hair roots, where even the most diligent preoperation scrubbing couldn't get them off. Over each square inch of the earth's surface, somewhere in the column of air that rises up to outer space, there are more than 4,000 dangerous, viable microbes — 12 hundred billion over a city the size of Chicago, or 400,000 for every citizen.

A lot of these microbes are quickly zapped into oblivion by direct rays from the sun, but colored ones just reflect this ultraviolet danger, and all of them can take refuge in the darkened shelter of a cloud, the shadow of a skyscraper, or just a darkened hallway, there to lurk like the ultimate city predator and swoop down when the air so moves them. Fifteen different kinds of virus capable of causing the common cold alone have been measured hovering over London streets. It takes only a few hours to breathe a room-sized volume of air through your lungs, and in that amount of city air are also likely to be at one time or another the patiently floating microbes that can cause influenza, meningitis, pulmonary tuberculosis, measles, diphtheria, pneumonia, other infections, and even gas gangrene. Not all bacteria are bad of course. One sort enters the intestine of all newborn babies, even in isolated wards, and there produces vitamin K, the clotting agent, as a digestive product. No one could live without that.

Air is such an extraordinary disperser of these tiny organisms that whole colonies of germs from floating scum in some lake or ocean can be shaken loose by a single thudding raindrop, and float a thousand miles without losing their potency. Some enterprising students at McGill University, for example, were able to find on the roof of their dormitory a colony of an algae-resembling fungus that usually lives in the open ocean and must have floated the many hundred miles inland to this new home in Montreal. A single raindrop landing in a fungus or bacteria-covered twig can shatter into some 2,600 droplets, each one of which will be likely to carry some of the infectious agents with them as they scatter.

But there is one way of germ dispersal mightier than all of these. This is the sneeze. It is such a good way to launch bacteria or viruses into the air that some scientists speculate that certain microbes evolved the irritant ability to make people sneeze just for this purpose. A good sneeze will send out one million droplets, each one loaded with a range of microbes that can cause havoc wherever they land. While most of us are immune to many germs in our own mouths, we're usually less immune to those in others'. In a sneeze, the germs shoot out at 100-plus miles per hour, way faster than even a cork from the most overpressurized of champagne bottles. Within a quarter second of their sudden arrival in the outside world most of the expelled droplets will evaporate into something resembling a tiny mucus-coated pinhead, with the ever-ready viruses or bacteria now nicely protected inside. At their small size these shrunken messengers are buoyed up in the air as a boat is pushed up in water and can float, adrift, for weeks.

Even ordinary conversation will produce great hordes of these infectious expectorates, although here a curious distinction opens up. Not all words are equally loaded. Since most of the germs in our mouths slosh about in the front, under the tongue's tip and on the sloping wall of the lower incisors, the sounds that scoop them up and spit them out most efficiently are the plosive consonants *f, p, t, d,* and the sibilant *s.* The sounds of *r, q, e,* and hard *c* are safe. An onslaught of "cool jive having rare eloquence," rich in vowels and palate-backers, is

Exhaled air, photographed from the convection patterns it produces. This is ordinary breathing; in vigorous talking germ-laden exhaust flow goes many times further.

likely to be almost antiseptic; half an hour of "steady discourse es-
chewing pat display," heavy on the spittle-splatting *s*'s and *t*'s can produce over 1,000 germ-laden droplets in half an hour.

Resistance

With all these germs so constantly spread about, it seems incredible
that any of us should survive. Left to themselves once inside the body,
these germs can multiply and produce all the disagreeable signs of
pneumonia or flu, tuberculosis or scarlet fever. The key point is that
the intruders are not left to themselves. Although by itself the body
may have a hard time to take care of the vast hordes of germs than can
come in during an unsterile operation, it can do quite a lot against the
accidental, barely noticed infections that happen all the time. It's because of these automatic responses that they stay barely noticed.

The first defense is a warning signal sent out by chemicals and from
the first cells to be attacked. Cells that are about to die after microbes
have abused their hospitality manage to fling out an ammonialike substance that splashes onto unsuspecting tissues nearby. This is the much
publicized histamine, a name which sensibly enough comes from the
Latin for "ammonia" and the Greek for "tissues." Most of it just lands
uselessly on the side of the still unsuspecting neighboring cells, which
are themselves liable for attack by the same kind of microbe that destroyed the last ones. But as the attackers blindly fumble their way
over to these fresh cells, some of the histamine dribbles onto one of
the many ultraminiaturized blood vessels that weave throughout all
tissues.

The slightest drop that touches the vessel shocks it like an electric
blast. The minuscule blood vessel swells out in a rush to up to twice
its original size and, in the growing, opens gaps in its normally tightly
sealed walls. From those gaps, like the ideal U.S. cavalry, comes the
sole substance that can rescue the nearby cells, a special protein-
loaded fluid, always on store in the bloodstream. Shooting from
hundreds of tiny crevices and fissures, it catches the attacking microbes before they've gotten out of the region and douses them with
specially shaped proteins and other molecules designed to smother
them dead.

Also leaping out of the blood vessel walls are hordes of rapacious
white blood cells. Some bend over double to fit throughout the small
crevices in the blood vessel walls then make it the rest of the way to
the infecting bacteria in an amoebalike series of slithery stretches. The
white cell pushes out a part of itself a tiny fraction of a millimeter forward, then sloshes the rest of its body into that forward extended
bump. Reaching a group of bacteria it stretches around them, gurgling
as it moves, and envelops them whole.

Fifteen or more distinct bacteria can end up imprisoned within just
one of these sockhop-crawling white cells. The bacteria are likely to
remain alive for a while in there, struggling and squirting out all the
poisonous toxins they can as a last desperate resort, but finally the
persistent, all-enveloping attack marshmallow dissolves them dead.

The blood vessel is not done yet. It has still another defender to

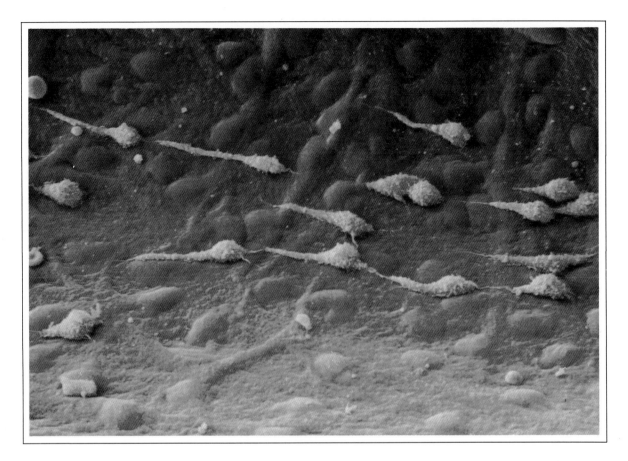

White blood cells speedily crawling along the inside of a blood vessel. These defensive cells have floated to the inner surface of the blood vessel from their usual position in the middle of the rushing blood flow. Their hasty migration on the wall can be followed by a burrow right through it, taking them to sites of inflammation in surrounding tissue where they can collect and destroy infectious intruders.

spring forth. These are the antibodies, and once they come out they're merciless, swarming on and obliterating any sample that they come across of the invading germ they're programmed for. The only problem is that they can take a while to get to it. It's an instructive moral tale to see why.

Antibodies are made of four toughened chains, which have been wickedly braided together. At one end of the chains are special recognition sites, shaped just right to dig into and attack the unwanted intruder. That's good. But at the other end of the antibody's chains — and this is what's less good — there are special sticky proteins that mire the antibody down in the outer wall of the first cell it touches.

What this means is that the antibody heads out, sharp chains swinging, but before it's really gotten anywhere it becomes caught tight in the outer edge of the very cell that produced it. And to make things worse, the intruding microbe that all the fuss is about is likely to be hovering, infuriatingly, right there, just out of reach. The antibody struggles to get at it, strains and pulls and quivers, but only succeeds in getting itself mired deeper. Righteous virtue has been taunted; evil remains scot-free.

It seems cruel for the ridiculously stuck antibody, but those furious flailings turn out to be precisely what provides the solution. The invader is soon to get its due. For the very struggles of the antibody in the wall of its production cell set off a series of construction steps deep inside that cell which lead to the production of new antibodies; antibodies that this time around will not have any of those task-stifling sticky bits to hold them back.

In less than a second these avengers hurtle out of their production cell, rise up past the mired-in earlier antibodies without even a pause, and home in directly on the still so insouciant invader germ. Whoosh whoosh whoosh! The new antibodies slam onto the culprit with their fresh chains swirling. They hack at it with the equivalent of deep gouging blows, and soon finish the miserable invader off.

The antibody's attack — once started — is as lethal as this because just the right design of antibody has been provided to attack whatever germ it was that approached. To keep this up though, to cover all the different invaders there might be in an 80-year life with eager invaders creeping in almost every day, your body needs to keep a gigantic supply of possible antibodies on hand. Carrying them all around preformed wouldn't be any good though — they would take up too much space inside you. The way the body solves this space problem, is to jumble and shuffle the heredity-controlling DNA strands inside the special cells that make the antibodies, jumbling them so efficiently that a stretch of only 300 or so gene units on the DNA can be permuted into over 18 billion different kinds of antibody defenders.

That's why the first antibodies to get made stop there on the edge of the cell. If the antibody isn't right for the current invader it will just sit there placidly and not cause more space-consuming copies of itself to be made. But if it is the right design for that invader, if its chains twist and scrape and lunge to get at it, then the cell it's based in will make more of just the right kind. Kicking up a fuss is what antibodies are all about.

This battle takes place on a highly miniaturized scale. If the invading germs are few and the defenders quick, it will stay on that scale and you will never know that it happened. But if the errantly entering microbes get a moment of uninterrupted action, enough to destroy a few hundred thousand living cells, then the possessor of the body is likely to be aware of the ensuing battle. He will commemorate the titanic struggle inside to lodge them from their stronghold by feeling slightly tired for a minute or two.

If more intruders were there at the start of the attack, he might even feel more tired. This time a real humdinger of a defensive battle will take place, stretching over distances unfathomable to the cells and antibodies involved. To us this will be as evident as the inflammation stretching for an inch or two around a slight bruise or cut. The redness and heat we notice are the dim sight we get through the skin's surface of the tiny blood vessels responding to the histamine's triggering order to enlarge. Any swelling around the cut will be the result of the torrents of fluid these ultratiny circulation channels spewed out to obliterate the invaders. If the battle is especially hard fought, enough dead invaders and defenders lying on the site build up into the pile that we generally refer to as pus.

Normally that's as far as it gets. The body's defenses clobber the gate-crashers before they can do enough damage for us to notice anything more. A lot more germs than usual have to get in for their unpleasant attacks to reach a level that these automatic defenses can't handle. Sometimes, though, we make it easier for the microbes. Staying up too late, grabbing junk-food snacks on the run, and all the other violations of Puritan ethics which make life so pleasant make it a lot harder for the body to clobber the quickly multiplying colonies of germs that get started in us through the frequent unseen infections. With not enough fresh proteins and other materials circulating in the body, there won't be the right ingredients to make a nice strong burst of histamine for invaded cells to squirt out to warn the adjacent blood vessels. A more watered-down warning will come out instead, and a delay of even two minutes before it starts the process going will let the germs increase through several generations of rambunctious reproduction.

That's a major reason why you're liable to get a cold or sore throat the day after staying up late: your enfeebled, watered-down germ defenses are no match for the flocks of infectious germs you continue to pick up as they float loose in the air, or even from any of the droplet-spewing plosive consonants hurtling your way. It's almost as if the feeling of general tiredness is as much a cause of the disease as the germs themselves. This point was flamboyantly demonstrated by one Max von Pettenkoffer, a 74-year-old Bavarian professor who in 1900 held up to his university lecture-hall audience a glass filled with millions of dangerous cholera bacilli, and then before their astonished cries swigged it down. He lived, and suffered no ill effects showing that anyone in as good a shape as he was should have nothing to fear from an infection that would certainly kill an enfeebled slum dweller. (Though most of the bacteria he drank were probably dead.) Everyone is likely to have a good range of infections in their body at all time — many of us have unpleasant polio germs in our stomachs, for instance. It's what our defenses are able to do against those infections that count.

The true solution to infections that somehow grow past the point where the body can handle them came only with the development of antibiotics around the time of World War II. There had been various "miracle" drugs before that, but most of these worked more by suggestion than science. The grand exception was quinine, produced from the bark of a Peruvian tree, which was a near-certain cure for malaria and related fevers — and used in the seventeenth-century by Spanish Jesuits to cure Catholics, and never Protestants.

Modern antibiotics have been equally exotic in origin. Penicillin is the best known: discovered when some culture droplets from Alexander Fleming's nose became contaminated by a bread mold, which stopped the bacteria growing — a fortuitous accident helped by overcrowding in his tiny laboratory following funding limitations. The paper announcing it was barely read for fifteen years, and taken up again only at the start of World War II by Howard Florey at Oxford. He was unable to produce it in England during the Blitz, and equally unable to convince the major American pharmaceutical houses to look at it, he had to get it manufactured by a range of small back-street chemists in central Illinois. Once recognized, penicillin was produced in mass quantities to save the lives of thousands of Allied servicemen. It was traded on the black market in postwar Europe, where the results were vast profits for those who adulterated it and horrid deaths for trusting children whose parents bought the adulterated version. Finally it was produced in quantity by the big pharmaceutical firms after 1948, heralding a chain of antibiotic developments that hasn't stopped.

At first all these new drugs had a near-total success in zapping the germs that caused a tremendous range of illnesses. They worked even when the germs had multiplied to a level much greater than the body could handle by itself. Meningitis, pneumonia, syphilis, and typhoid fever were no longer to be feared, much less the sore throats that once so scarily developed into rheumatic fever and heart disease. In fact, the new antibiotics were so effective that pharmaceutical companies were given carte blanche to promote their marvelous properties. This was a big mistake.

These drugs got their power by being specifically shaped to kill microbes in a very particular way, perhaps by blocking the way the microbes built a certain part of their wall, perhaps by stopping up the tiny paths by which they brought in food. Because of the drugs' specificity, any bug that was just slightly different from it would escape unscathed from the antibiotic blast that killed all the others. This meant that the odd bug would get a free field in which to reproduce and grow within the body that thought it had been cured by the antibiotic.

Now, one measly bug growing in a body is likely to get clobbered by the natural defenses, but not, perhaps, before it's been able to reproduce a few times and scatter a couple of its offspring to nearby bodies. If this were ever to go on enough, after a while there would be enough of the resistant kinds of microbes around to produce diseases that looked just like the nasty ones antibiotics were ideal for, but, since they were produced by the half-brothers, as it were, of the original bugs, the antibiotics would have no effect. This point has already been reached. In 1974 in Mexico 14,000 people died in a very unpleasant way from typhoid fever, a disease that is usually easily cured by the antibiotic called chloramphenicol.

Here's where the pharmaceutical companies figure. The standard drug reference book American doctors use says that chloramphenicol should never be used for "trivial infections." It also warns about possible side effects. But in Latin American countries, including Mexico, the standard drug prescription book given to doctors by the pharmaceutical companies for decades says that chloramphenicol is wonderful for a whole range of minor problems, including tonsilitis, sore throat, and skin sores. Mexican doctors obligingly did what they were told, with the result that bugs that were slightly abnormal, and so resistant to chloramphenicol, have had a clear field in which to grow. When they suddenly flashed to epidemic proportions, as in 1974, the overused chloramphenicol could do nothing against it.

It's happening in the United States, too. Doctors often give you a dose of antibiotics to shake off a cold, instead of recommending a few days in bed. This is an utter waste of time, for colds are produced by viruses, and, as we've seen, antibiotics have no effect on them. What the shot does is give any slightly deformed germs a clear field to grow inside you and, quite likely, spread throughout the population.

These resistant germs don't just take up nest in new bodies which would be only a bit unfortunate in itself, but they go on to transfer their invulnerability to other, potentially very grave, microorganisms that they find. Already there are several strains of gonorrhea that are immune to most usual antibiotics. Getting an antibiotic shot as a precaution before suspect sexual relations means only that when these gonorrhea germs arrive in the bloodstream, they will adjust to the level of antibiotics there and even produce enzymes to break down the antibiotics so that they can grow more readily. Worse, a number of American children have already come down with a new type of meningitis. Because of the immunity picked up from the wholesale use of antibiotics elsewhere, these children cannot be cured.

Parasites

If researchers do not keep up with the mutations and changes in microbes, and steadily produce ever newer, better antibiotics, we might well return to the situation where almost everyone has a bit of an infection in them that's not fatal, but just enough of a drain to keep them regularly under par. This is still the situation in many poor countries today, and a chief culprit there, as it was for almost all our ancestors, is worms.

There are roundworms, tapeworms, hookworms, and whiplike worms, all floating about in a variety of unpleasant places and all ready to crawl into a comforting and nourishing body as tiny specks, there to grow and grow and grow up to a length that can reach, in an average 5½-foot human, 32 feet. In many tropical countries baby worms will be writhing eagerly in almost all swampy paddies, or in the dust of all village paths, and will easily slip in through cracks in the skin on the feet of the largely barefoot population. Once inside, they do exactly what the appellation "parasite" suggests, for the name is from the Greek for "one who eats at somebody else's table." In this case the table is not a courtly dining hall but the soft and squelching tasty walls of the infected one's stomach or intestines.

The pork tapeworm shows how it's done. It first presents itself as a tasty pinhead-sized morsel and then with almost no brain apparent to the naked eye or a microscope proceeds to be eaten by a pig, burrow into its flesh, turn into an opalescent larva there, survive the crunching and chomping in the mouth of a human attracted to partially cooked pork, and settle down in his intestine. Then in a burst of cooperative endeavor that would put a precision flying team to shame, the larvae form a long chain in which the first ones in line grow claws to hook onto the intestine, the second ones impregnate the third, the third hold on tight while developing eggs, and the members of the fourth and final group, way back at the tail-end of this chain that can be up to 22 feet long, pop out their fully formed worm eggs to journey into the world and start the what is to them miracle of life again. Yuck! Such intestinal inhabitants are common in China, India, and much of Latin

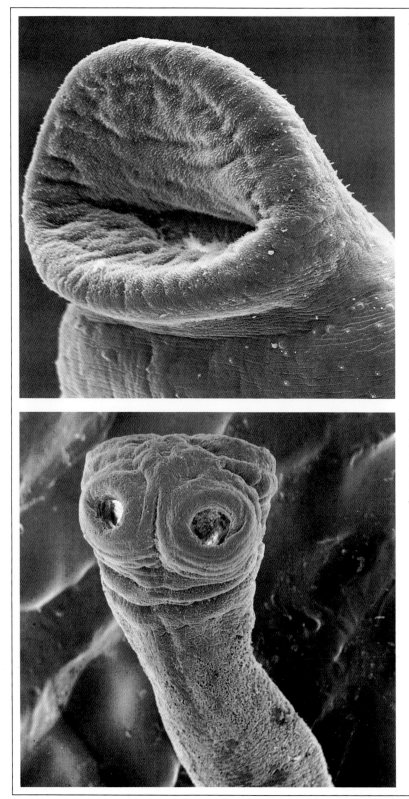

Mouth of schistosomiasis — one of the most common bodily parasites, and endemic in many millions of people in poor countries. A voracious eater, it saps its victim of energy.

Head of intestinal tapeworm that can stretch for several feet. Essentially one long feeding tube; through what appear to be eyes one can see start of digestive channel. Acquired from poorly prepared pork.

America, steadily draining their hosts and keeping them in an exhausted lethargy — not at all right for the entrepreneurial go-get-'em spirit that development capitulation requires.

It's perhaps understandable that one of the first biologists to discover the path of such parasites, Louis Pasteur, took to examining the food he was served at friends' houses, with a portable microscope he carried tucked under his coat to see if it was safe to eat. Even farther back, the authors of the book of Leviticus may well have had the tapeworm in mind when they banned pork products to all who wished to be Jews (a sensible precaution in the days before refrigerators, which was mimicked by Mohammed in the Koran).

This does not however guarantee complete immunity to those who follow the kosher laws. For tapeworms can also grow in other ways, and in one sneaky guise they home in on probably the most thoroughly kosher subgroup in North America: Jewish grandmothers.

The infestation comes in the course of duty: the tasting of not-yet-cooked gefilte fish to be sure it'll be up to standards (for since when is just ordinary gefilte fish good enough?). Alas for diligence, in uncooked gefilte fish there are sometimes uncooked bits of whitefish, and these in turn contain bits of water fleas that contain bits of fish tapeworm eggs. Diphylobothrium lotum. The start of this chain is sometimes at near-epidemic proportions in the whitefish population off the eastern coast of the U.S. and Canada, and from the original egg, through flea, fish and tasting, a tapeworm up to 32 feet long can develop in the unsuspecting final consumer. Efficient detective work on the ensuing "gefilte fish gastritis" is what brought this chain to light in the final place, and it can be ended painlessly by a quininelike drug and a stern injunction for no more nibbles — not even a small one.

These worms though unpleasant are no more parasitic than the much tinier bacteria and viruses we constantly harbor. And just as many of the tapeworms can take up their home and stay in the intestine, slightly lowering iron absorption each year but rarely going far enough to be fatal, so the microorganisms that parallel them in the parasitic life also rarely go so far as to kill their host. Any bugs that did always kill the creatures they lived in would soon run out of places to inhabit. It's the same principle behind the IRS not slapping a 100 percent tax on you, for even though they might like to, that would leave nothing at all for their taking next year. That's why most of the numerous infections we get will go away by themselves in a while, for the bugs that make them are self-controlled enough so that the body can keep them under control. Usually. It's incredible how these parasites have evolved to carry on their host-inhabiting life-style. One parasitic worm that goes for shrimp lodges its eggs in the shrimp's brain, that disorients the shrimp so much that it will scamper *towards* its archenemy the seagull, whenever that predator appears. While a nasty death for the shrimp, it does up the odds on the original worm's eggs getting eaten by the gull and transported far and wide for future proliferation: typical behavior for a parasite!

Humans can suffer the same evolutionary takeovers from parasites too. In Central Africa, eggs of the worm that causes bilharzia will often

"Little fleas have littler fleas upon their backs to bite 'em, and little ones have littler yet and so on ad infinitum." The general principle of symbiotic parasitism — here illustrated to perfection in these micrographs of a hedgehog flea with its mites.

pop out from their resting place in stagnant water only between 12 noon and 2 P.M. — just when the local tribespeople are most likely to come down to the water for a wash. Little do they know what's lying there in wait! Even more insidious are germs that cause elephantiasis. Swirling through an unlucky human's bloodstream, they have been seen to lie quiet for all the day until, just before sunset, they swarm up to the skin's surface in great numbers — just in time for pickup by their nocturnal mosquito carriers. What bouts of illness this might cause in the human host are totally beyond the concern of the so appointment-conscious bugs.

Burns

Infections are one side of the great illness divide. Collisions with the outside world are the other. They are much more certain to push you into the category of being ill, and they include bruises, bangs, wallops, cuts, scrapes, and, above all, as the insistence that can serve to cover the misery of all the rest, burns. No one with the exception of the G. Gordon Liddys of this world (who reportedly held his clenched fist in a candle flame to prove his fortitude to his CIA superior) wishes to be burned and there is good reason for this.

Even a small burn, such as comes from brushing one's foot against a hot motorcycle exhaust pipe, produces a disheartening amount of local damage. To see just how much, imagine a clock that chimes each thousandth of a second. Before the first chime is up, the skin will have taken in so much heat from the burning object that it vaporizes where it was touched. In the next five chimes the heat will spread to the small cavities of fat deposits that lay just under the surface, and will heat them like oil trapped in a tightly sealed pot. A closeup look would show these fat globules first begin to shake and shimmy, then in an incredible process, actually heat to boiling, glow, and steam away. In the next twenty chimes of our fast-sounding clock the thick web of nerve cells lying under the skin come under attack. The heat will force away their sodium and calcium fluids, melt the substances in their tubular walls, and collapse them into fused wrecks.

Any of the čapillary blood vessels where the burning object touched would have been severed open in the first chime. In the half-second after the nerve cells die, the blood vessels farther away will have part of their walls melted away or will get small tears from the quick twisting of local muscle fibers that have themselves been flung away by the first damage, and scrape against the walls of these farther blood vessels as they writhe. A burn the size of a thumbnail can destroy in that tiny area 8 yards of nerve cells, 300 sweat glands, and 6 feet of blood vessels.

All this is only the start of the problem. While the actual burning is taking place, the hot object against the skin heats the surrounding air and sends it billowing away in rushing convection currents. Once the offending object has moved away — or more likely has been leaped away from — then the thousands upon thousands of the bacteria that are continually floating in the air will have one of their best chances ever to get at their ideal source of food and habitation: an open hole

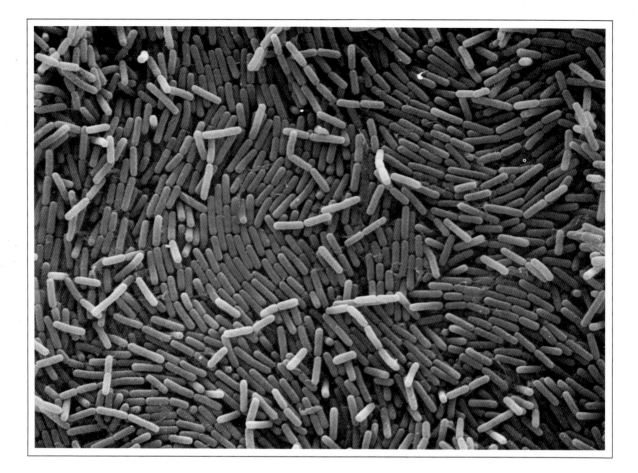

in the skin through which to enter the body's bloodstream, lymph system, and, from there, all the appetizing tissues, muscles, and organs in sight.

The danger of this happening is one of the main reasons why the skin has evolved as an impermeable sheath. The fourteen-odd pounds of skin each one of us has are almost entirely waterproof, and produce sweat that is mildly antiseptic. But a burn's suddenly produced entrance can lead to populations of millions of potentially disease-producing microbes that have landed out of the sky in the first two to five seconds after the burn and have begun to reproduce inside.

Nor does the danger go one way only. Even a burn a quarter of an inch across seems awesome to the body fluids inside, and so out they pour. The rush can include the plasma that carries the blood, the lymph fluid that internal tissues need to wallow in, as well as red cells, white cells, and the like. In a major burn this loss of fluids is the prime worry, for in half a minute it can cause the smallest arteries to collapse somewhat through the lack of anything inside to hold them open (though they've some elastic in their walls for this), largely through the sheer loss of volume, rendering ineffective the body's last-ditch attempt to save itself through shunting the remaining fluid from the peripheral circulation to the central to keep the brain and kidneys going.

Swirling pool of germs that can quickly fill in burn once the normally resistant skin layer is destroyed.

This sends the blood pressure plummeting down to zero, and from that there is no recovery.

As rough-weather specialists know, the body can survive much more easily without food than without water. The usual survival guides are three minutes without oxygen, three days without water, or three weeks without food. More precisely, a loss of water equal to 1 percent of the body's weight, three pints or so, produces a noticeable thirst; a 5 percent loss produces collapse and a 10 percent loss — as might happen in a major burn — is fatal.

These consequences sound so unfortunate that it's surprising anyone south of the Arctic has survived the discovery of fire. But in fact most burns heal without problem, and even people who have been burned through to the layer at the very bottom of the skin, and over 65 percent of their bodies, have recovered when in the best medical care, using computer-assisted nursing. What's going on to account for this disparity? A lot, and most of it without our realizing it. As is so often the case, a little etymology comes in handy.

The ancient Greek word for "burning" was *kaustikos,* a pronunciation that has lingered with little change to become the modern "caustic," which still means a burning substance, or an abrasive or cutting remark. Athens had a slave population many times greater than its number of free citizens. Many of these slaves had been captured in foreign wars and had to be marked as a hedge against their escaping. Branding was often used in a way that would not harm the new slaves. And what was the word used to describe this noninjurious branding iron? It was *kauterion,* just a variant on the *kaustikos* word for "burning."

No new word was needed, for the Greeks realized that a brief, intense burn could heal itself, producing no dangerous outpouring of bodily fluids. This works because the heat that forms the extreme limit of a burn in the body also serves to close up small blood vessels, seal leaks of plasma and lymph, and generally fuse together any substances to rush out. This closing is the chief spontaneous aid to keeping the body fluids in, and dangerous microbes out.

The Greek knowledge of how this worked made it across the Aegean with little loss to the Romans but in the ensuing Dark Ages it became curiously distorted. By the late Middle Ages the valid knowledge that a brief intense heat could cause blood vessels to fuse together and to stop bleeding from a wound had become twisted into the remarkable notion that the best thing to do to someone suffering from a gaping wound was to pour boiling oil on it — in fact the worst possible treatment for a man suffering from one of the messy battle wounds that the pikes, boulders and spiked armour of that God-fearing time produced.

The first experiment that put the boiling oil treatment to the test and so set the stage for all of modern experimental medicine, was unwittingly organized by a twenty-six-year-old Gallic barber and part-time surgeon named Ambroise Paré, in 1536 on the field of a particularly gory battle where he was treating the wounded. After the senior doctors stopped work that night when they ran out of the usual boiling

oil to apply, Paré searched out an alternate dressing, and ended up
with a barrel full of egg yolks, rose perfume, and turpentine. This salve
he proceeded to slop onto the wounds of the groaning soldiers —
who probably thought their luck had run out when they didn't get a
chance to be treated with boiling oil — and it worked.

(Paré himself, a Protestant, ended up as head surgeon to the French
king and in recompense survived the night of the St. Bartholomew's
Day massacre many years later — tucked away in the back of a closet
in the royal dressing room. His new treatment was a welcome alternative to the leaders of church and state at the time, who would otherwise have to suffer the old boiling oil regime.)

The body's other safety measures against damage from a burn begin almost as fast as the cauterizing effect. Plasma and other fluids that
leak out of the damaged blood vessels pool in barely visible lakes on
the surface. In contact with the air's oxygen they change into a brew
that serves as an excellent barrier against any malicious microbes that
might wish to land there. (They also help block the nasty toxins that
some of the bacteria gush out.) Inside the plasma fluid are lots of
grainlike black objects — 40 to 60 million of them in a 2 cc drop of
blood. They were originally spewed out by giant production centers in
the body's bone marrow, and now have a special task of the heart of
the dike-blocking task.

For while the black grains are piling up at the burn, yet another
protein in the blood plasma is being converted into rubbery long fibers at the burn site. These fibers loop into a tight meshwork, and inside that netting the black grains and certain other blood cells get
caught and so block the dike! Three hours after the skin broke it can
be complete, and four or five days later the fibers will have become
mighty collagen — which, once formed, has a tensile strength, resistance to a tug, not far from that of steel. The resultant solid covering
over the wound is called a scab, possibly from the Old Norse *skabb,*
meaning "roof."

Any underlying nerves that were only partially fused will begin to
regenerate, and the underlying muscle, as well as the little sacs for oil
and perspiration fluid will begin to grow back, edging out from the
nearest undamaged tissue. If the burn is too deep, however, there will
be no nearby undamaged tissue to serve as a base for this growth, and
any skin that does manage to sprout its way back there will lack the
thick and malleable underlying layer of muscle that most skin has cut
deep enough to destroy the layer just under the top of the skin will
produce this featureless spreading of skin across the gap. That's how
scars are formed. The word *scar* itself demonstrates the folk experience that burns are often at the cause, for it came from the Greek
word for fireplace; in the centuries before stoves and central heating
this was the place where most burns were sure to take place.

Large areas of burn scarring look thin, stretched, and taut, which is
exactly how they feel to the unfortunate ones who have them. Trans-
planted skin from other persons usually is rejected by the burn victim's immune system, though transplants from other members of the
family do better, and there have been instances of quite noble broth-

Another type of infection that can speedily build up on a temporarily open burn: an omnipresent fungus, related to the mushroom, and far too small to ever see with the naked eye, as seen under an electron microscope.

ers and sisters who have chosen to risk extensive permanent scarring to give large quantities of their skin to severely burned siblings. Skin from elsewhere on the burned person's body is even better, for then the body's ever-vigilant antibodies have no incentive to slough it off, but in precisely those cases where it is most needed there is not enough elsewhere to give.

Animal skin donations have been tried, and it has turned out that the common pig shares enough chemical traits with humans to be a good choice. In several cases a rabbi has been contacted to see if he would give his blessings for a badly burned Orthodox Jew to get this salvation from the most strictly forbidden of nonkosher animals. This blessing is always given, as the Talmud has a clear, and benevolent interpretation here.

Automatic attempts to reverse the damage caused by a burn can come into effect only once the burning has stopped. It is on this scientific principle that hardworking chemists at the Dow Chemical Company were able to perfect a new weapon of war: napalm. This patented discovery, which even such nonsqueamish individuals as Winston Churchill and the commandant of the U.S. Marine Corps have recommended outlawing, works by treating normal petroleum until it

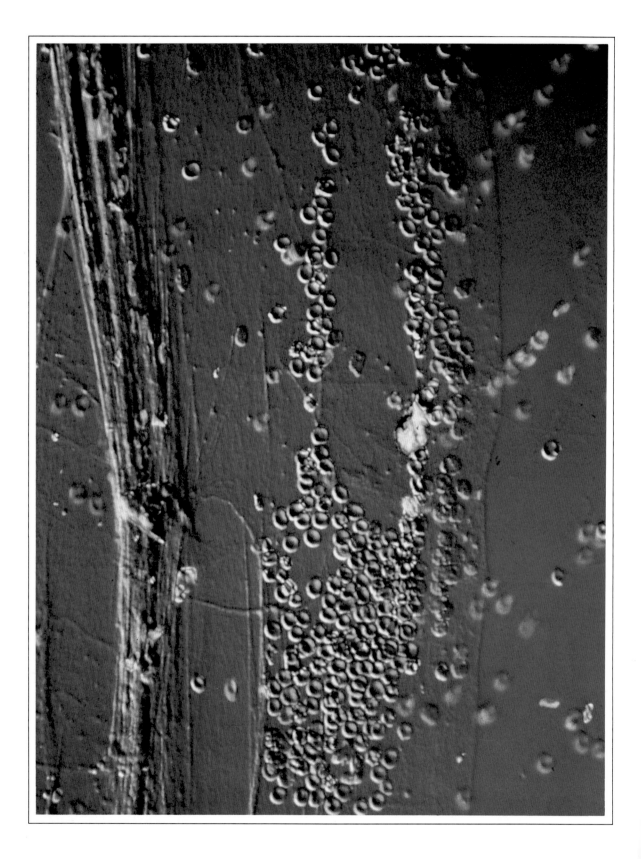

becomes thick and sticky. The resultant mixture is wrapped around a long electronic fuse and poured into a thin metal house, aerodynamically shaped to drop smoothly from the bay of a soaring bomber.

As napalm was used in Vietnam, the fuse ignites the petroleum when the gracefully falling bomb is a few hundred feet above the ground. The igniting petroleum explodes with enough force to smash apart the thin metal casing and send bits of burning petroleum in all directions. If it were simply burning petroleum, it would disperse into tiny drops and the fire would go out before any reached the ground. But as the chemists mixed in a sticky ingredient, napalm always came down in good-sized globs. Anything it landed on, be it house, beast, or man, it would stick to, while the fire raged.

In the course of the war, euphemistically named "quality-control specialists" found that the victims were often able to brush off the flaming globules before too much damage had been done. The American designers soon came up with a revised sticky ingredient that melted farther into the skin and so could not be brushed off. Even that wasn't the end of it. The quality-control people reported that some of the Viet Cong it was dropped on had apparently been able to put out the fire by leaping into streams. To this the Dupont chemicals replied with what was perhaps their ultimate refinement: the mixture of added phosphorus (which, as the medical alchemists found, burns under water) to the sticky mess inside the bomb case. Napalm so refined will burn on contact, burn if the sufferer tries to rub it off, and will continue to burn even if the victim were to be totally immersed in water. While it burns, the body's natural defenses — the fibers, black grains, and the rest — can do little to help.

A totally unexpected consequence was the development of computer-controlled burn units, originally for the large number of American servicemen who kept on getting accidentally doused with the stuff.

REFLEX ARC Because burning has such dreadful consequences, there has been a marked evolutionary tendency in all land creatures to avoid hot objects. Water dwellers are usually spared such concerns, which is why a dolphin coasting on the surface will have no trepidation about heading straight for a lit match. In humans this evolution has developed into what is the very fastest reflex in the body. Accidentally graze your hand against a boiling-hot teakettle, and the first thing you're aware of is that you have jerked your hand away. The sensation of burning comes only a few thousandths of a second after that.

That's fast. The reason is that a lot of burning can go on in the time it would take for a heat-sensitive nerve in a finger to be triggered by the hot kettle, send a message to the brain, and get a reply back to the arm and hand muscles, causing them to move away. These messages course along the nerve pathways at about 50 mph. At that speed the round trip would take enough time for a lot of powerful burning to occur.

To get the exposed finger away from the danger before incineration sets in, the nerve that registers the pain is rerouted when it reaches the spinal cord. The spinal cord can be thought of as a leftover and

Scab forming, as fibrous strands created from substance in blood plasma weave net that catches blood cells. Vertical sheaf on left are well formed, the weaving strands in the middle are just beginning.

primitive predecessor of the brain, and usually it just passes upward
the messages that come to it. But when such a crucial message as "I'm
burning" flicks in, its simple and lumbering processing capacities stop
lumbering and snap into action.

The message describing the burning is relayed upward to brain
headquarters, true, but at the same time an emergency message to the
muscles controlling the exposed limb is sent by what seems the spinal
cord's own initiative (and is actually just useful wiring!), telling them
to tighten immediately. That starts the limb on a trajectory away from
the source of the burning much more quickly than if the brain was
used. In most children that reflex pulls the arm away from an unsuspected sizzling object faster than the best prizefighter can throw his
fist in a jab — even compared with laboratory measurements taken on
Muhammed Ali when he was in his prime. On this reflex arrangement,
even a pudgy accountant will outdo the Greatest. Not bad, and so surprising in fact that two generations of neural investigators never found
out how this happened.

Credit for the discovery goes to Charles Scott Sherrington, a delightfully eccentric researcher working at Oxford University early in
this century. In a series of experiments on dogs, Sherrington dissected
a number of animals and stimulated their nerves until he was convinced he had found the nerves that led away from a suddenly excited
heat-sensor, the place in the spine that decided to send the first return
message, and the return nerve cells that carried this call. His name for
the whole system was "reflex arc."

As Sherrington described these reflexes, they fit quite neatly in with
the fashion of finding hidden mechanisms that was so widespread in
Europe in that rapidly industrializing time. Marx had found an all-per-
vading substance called "ideology" that he held fit into very precise
pathways; Freud was in the process of doing the same with the sex
drive and its precise channeling within the unconscious. Perhaps because Sherrington's ideas were so consistent with this fashion for preset pathways, they were picked up by a large group of researchers in
neighboring fields. The Russian Ivan Pavlov is the most famous, with
his demonstration that animals can be trained to salivate or to bark
even if the stimulus of the action has no relation to what would start
that action in the wild. Through the American psychologist John B.
Watson, this developed into the system of behaviorism and conditioned reflexes developed by unlovely B. F. Skinner at Harvard University. Modern concepts of how advertisers can best get across a message, how body language works, and why crime is high in slums
unfortunately all hinge on the notion of conditioned reflex. And that
came from Sherrington's study of why we leap away from a burning
object so quickly.

In conclusion, a visit to the Arizona desert. Not the scrubby hills
near the Colorado border, not even the gas-station-strewn cactus
growth that runs along the interstate, but the real sun-blasted stuff, way
out in the parched outskirts of Tucson, where the humidity is rarely
over 2 percent and it rains at most once every five years. This is the
site of the Davis-Monthan storage facility — "boneyard" in the trade —

where the U.S. air force has assembled the largest collection of out-of-service warplanes in the world. Multiengined B-52s are there, dozens of them, as are F-100s, propeller-driven transports, and hundreds of others, from the Korean War era on up.

There's no other place on the continent so well suited for the storage of mechanical equipment. There's no moisture to soak into the decades-old wirings, no freezing temperatures to crack engine cases. The planes seem to be in as good shape as they were when setting out on their last missions. *Seem to be.* For even out here, where storage conditions are perfect, none of the craft would be able to take off again without a thorough process of tuning up and cleaning.

There's likely to be dust that needs to be cleared from the ignition circuits, or turbine spark plugs that need to be refitted. With a crew of air force mechanics working together, it takes 48 to 72 hours for each aircraft. Once taken care of, the propellers could turn, the turbines whir, and the long-stilled planes could lumber off into the sky again. But without this last-minute work, they just wouldn't go.

In its own long storage ability the body does far, far better. All the natural mechanisms we've seen in this chapter, for pain, germs, and burns, last immensely better inside the warm, wet, jiggling body than any of the flying behemoths do in their perfect setting outside. A middle-aged college professor who perhaps last scorched a finger thirty years before, and has lived a life of the meekest self-preservation will unknowingly still have all his antiburn reflexes charged up and roaringly ready to go. He just needs accidentally to drape one hand on a boiling coffeepot, and, without a slip, without the slightest need for an overhaul or cleaning, the long-forgotten circuits flash into play.

The reflex heat receptors, which have waited their turn patiently and uncomplainingly for thirty years, shoot their message right to the spinal cord. The right prerigged circuits in the spinal cord just as quickly send off an immediate order to the hand muscles to pull back pronto. All the rest of the body's defenses speed back to work, too. With hardly a delay the pain signals are once again rounding their banked turn in the armpit on their way up to the brain. The white cells are speeding out from the small capillaries to gurgle their way onto any germs, and the right protein strands are getting ready to form their defensive scab over a burn. All work in the first few seconds, all without the slightest need for practice or refitting.

All the other pain and illness responses as well as the burn ones stay resting unnoticed in the body too. Those are the pathways and powers in us all the time just waiting to be called into action. Reflection on their rarely counted, but ever-able existence can do wonders to make you feel that you're never quite alone, whenever you're lucky enough to have your body along where you go.

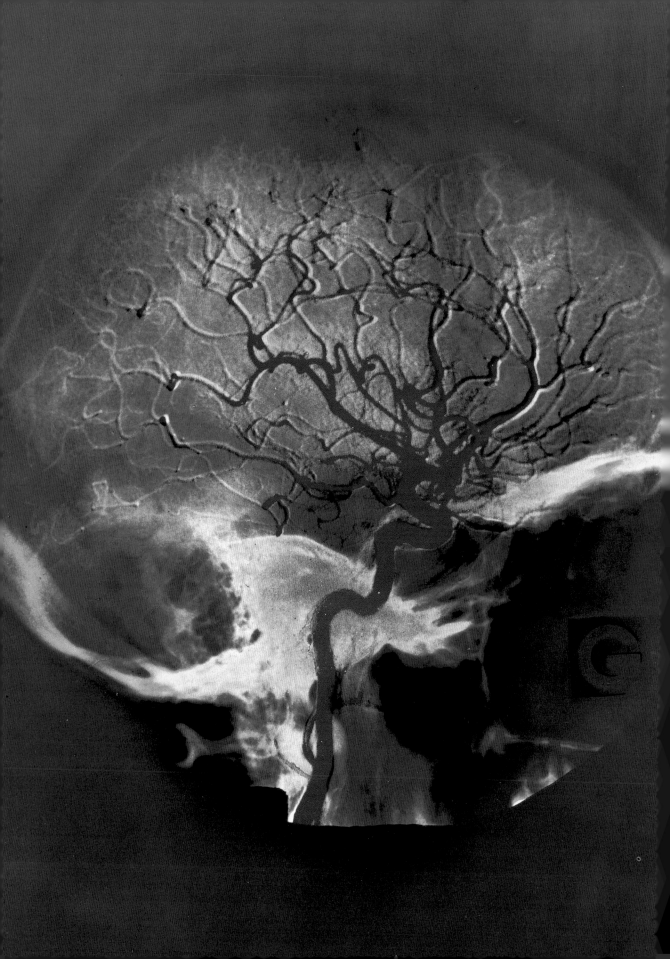

6
Stress and Worry

Coping Introduction: The Nail

Pity the poor nail-biter. He tugs, he pulls, he chomps and he crunches, but almost never does he win. Sunk in his debasing habit, primed to instantly stop if someone enters the room, why ever does he do it?

He does it because the phone rings too much. He does it because the elevator was late and the computer's out and the guy in the deli on the corner is probably going to make fun of his shirt again at lunch. He does it because there's too much going on in corporation America for his fragile superego to cope; he does it along with millions of other unsung nail-nibblers across the country: all hunched forward and littering floors with the fruits of their secret stress-caused shame.

And they do it because the human nail is a modified bone that hasn't quite evolved into a hair and is just PERFECT for making a little bit of that awful enveloping stress go away. Like this:

Early vertebrates were entirely covered in thick scales, as insects are today, but this direct inheritance we keep only in our lingering externals of an embracing skull and a surface-projecting collarbone. Early reptiles evolved to grow some of their scales thinner, then birds did it even more, and by man the result in most places has been just delicate fibers of hair.

Only our fingernails petered out a bit before the full way. They're no longer dinosaurlike scales, but they're not yet soft downy hair either. All human fingernails are made of protein, 96 percent pure (that's four times richer than prime-cut sirloin), and they divide into lengthwise strands that fibrously ooze out just under the cuticle at their base. That's where they grow in us, faster on the index finger

The incredible flow of blood into the brain. Note, however, the pain of headache occurs on the external blood vessels outside the brain.

than the thumb, quicker on the right hand than left (reverse for lefties); and edging out at an average rate of four inches a year, or enough to equal one's own height every seventeen years or so — say by the year 2000.

At first it would seem that the teeth should be able to tear off as much of these reptilian remnants on our fingertips as they want with no problem, since teeth are covered with an enamel coating that's the hardest material in the body and can be propelled by the jaw muscles to generate close to two hundred pounds of pressure per square inch. Chomp! It's enough to crack a good-sized chicken wing, so surely should do for the pitiful protein strands that make up the nails.

That, however, ignores the way nails writhe and twist when the cuspid crunch comes; the underlying battle is not really as mismatched as it seems. The nail's protein writhing diverts the toothly force down along the whole length of the nail, and leaves almost none to cut across sideways at the particular point of impact. Human nail has a resistance to fracture forty times greater than that of stone or brick because of this twist.

The simplest way out, of course, is for an attacking tooth to take one *quick* snap when the nail is not expecting it — before the recalcitrant writhing can begin. Why then do confirmed nail-biters forgo this opportunity for success, this single moment when a dry and most brittle possible nail has the greatest chance of succumbing to their oral advances? Why do they instead dribble and slobber over the objects of their attacks, rolling the nails and even whole fingertips around in their mouth, chomping a little bit here, gnawing a little bit there, but never going at it in the most scientific way that would win them their apparently greatly sought-for prize?

The answer must be that the nail is not the real object. Rather the pitiful goo-producing loll of the fingertip around the lips, gums, and teeth must be what provides the motivation. And there's support for this. One single roll of one nail inside the mouth sends electrical signals firing solidly in a wedge of brain tissue in the sensory cortex — the sense reception center of the brain — that stretches for 1¼ inches across, a target that's lots bigger than the mere ⅛ inch of brain tissue in the same sense reception center that is responsible for all stimuli from the genitals.

Oral satisfaction is what nail biting is about, all in the guise of an assault on these finger-crowning evolutionary holdovers of our outer skeletoned past. And that gets rid — at least for a time — of the stress of daily anguished life.

That's not the only means of coping with stress, of course; on the contrary at times it's almost impossible to keep the body from coming out with all sorts of defenses against the stress we insist on living with. Sometimes it's a simple flood of adrenaline, as we saw in the special case of fear and anger back in chapter two; sometimes it's more long-lasting or just plain bizarre, and can do things all the way in your bowels, or make you cry, or foul up your immune system, or just make you keyed up and a pain to be near.

The key notion is that no one can keep up a stressful life without

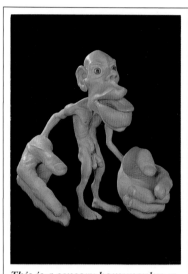

This is a sensory homunculus, a human figure distorted so that its parts are proportional to space the brain gives to sensations from different parts of the body — thus hands and tongues dwarf back and thighs, and rolling of fingers on tongue and lips in typical nail-biting produces more sensations than all the rest of the body, including unaroused genitals. Only the eyes, of course, are given more sensory input space in the brain — they would if sculpted to scale be larger than all the rest of the homunculus!

Human hair, showing its force-transmitting fibrous internal protein structure, shared with nails. If it looks like logs, that's because both grow outward from a central core.

having the body start on a vast number of contortions in its attempts to cope. And for one of the most pervasive stress defense of all, the contortions and changes start with a chemical once known as a bile solid, and renamed as the fancier choles(bile)-sterol (solid): cholesterol.

Automatic Coping with Stress: Hormones — Cholesterol, Cortisol, and Control

Cholesterol is sludgy and gooey and if you eat lots of butter it clogs up inside your heart and chokes your arteries and makes you die.

Well, not exactly. In excess there can be problems, but cholesterol normally is one of the key chemicals inside your body — at the core of the sex hormones, vitamin D growth, and is even 15 percent of the brain by dry weight. Cholesterol was first created by ancient bacteria that needed some way of keeping their slippery cell membranes rigid. Because of that origin hormones built up from cholesterol even today are incredibly good at crossing biological membranes.

These hormones are called steroids (chole-*sterol*) and are what East German female swimmers take to give them the vein-gnarled triceps that pull them to victory ahead of their dainty, merely feminine, American and other Western competition. The way steroids tie in with stress is that when you get too much aggravation, a potent hormone based on this membrane-cruising cholesterol soon comes pouring out from a storehouse down in your back.

The process is one the gentle reader should have some intimation of by now. Imagine that you're sitting at your desk, trying to write a report, after innumerable interruptions by the same wheedling assistant, when the send-off action happens — the assistant barges in with another needless request. All you'll feel is another burst of exasperation, but what will really happen is that the hypothalamus in your brain will be started into action once again. It won't let out with the LHRH that was behind sexual libido, nor will it start off commands that end with a simple adrenaline surge. No, this time it's a different triggering chemical that goes dribbling down into the pituitary, that tiny spheroid of a gland encased in the bony roof of your mouth. The pituitary swallows, rumbles, and then, ever so gradually, ever so gently, lets out with the juice that will cause the steroid stress hormone in your back to get to work.

In that juice are wispy, canoelike strands, each made of thirty-nine amino acids elegantly stitched together. Once they glide out there's no turning back, no quick changing of your mind to reverse the momentary burst of exasperation that started them on their way. What happens next is inexorable.

Silently into the bloodstream the thirty-nine-stitch canoes float. Some get rent apart by powerful enzymes at loose in the blood; others get lost over in crannies in the arms or feet and gradually are dissolved apart in the churning blood there. But some — and these are the ones that count — end up floating into a haven of a lagoon. This is the outer edge of the adrenal gland (adrenaline was made in the middle bit), perched up over the kidneys, and each one is filled with quiet blood canals, undisturbed, and unbuffeted. Those canals are lined with rows and rows of steroid hormone storage tanks, and nuzzling up against them is where the tiny pituitary gliders finally come to rest.

The time is two minutes since release from the brain. The situation is one that's about to explode.

Each storage tank has little handles on the outside that the pituitary messengers expertly berth on to. (They will work even if the canoes have been buffered, and a full third of their stitches fallen off.) That contact starts the storage cells to work. A little bit of the steroid hormone that was inside them starts to trickle out of the equivalent of faucets on their surface, then some fresh steroid hormone is made from the cholesterol that had been stored away inside, then some more steroid comes out and soon, moments after the contact started, the lagoon canals start filling up with steroid.

Within three more minutes they're full. Back in the office you might be leaning over your desk once again, maybe even humming, and most likely will have forgotten the irritation of a few minutes back. But since you did let yourself get irritated back then, there's no stopping the steroid hormone brimming up in those adrenal lagoons now. A bit more squirts out of the storage tanks and that's it: over the edge, down the canals, and out from the lagoons the whole buildup pours, tapping into the heavy blood by the kidneys and soon reaching every part of the body. Out there it raises blood sugar, speeds up brain firing, and rouses the muscles into a quivery readiness. If there's action called for you'll be ready for hours; if not the triggering stress will just have made you jittery and ill at ease.

All this followed inexorably from the moment the first pituitary canoes were released; all this works now because of the way the steroid hormone (the particular one we're following here is called cortisol) can weasel through the normally invincible cell walls that it hits. The principle it uses is the same as one a temperamental pianist might use who insists on fine-tuning the heat in an auditorium where he's going to give a concert that evening, but who is not allowed by the no-nonsense janitors to warble and scamper his way down the stairs into the basement and switch on the right furnace blowers himself.

A particularly ingenious pianist could outwit the janitors by flinging his arms over his piano on the stage, then declaiming in a loud voice that by union regulations all pianos have to be lowered down on the stage's hydraulic lift into their storage spot in the basement as soon as the afternoon practice is done.

Presented with that conundrum, the janitors are likely to give in, and slide the piano — with demented pianist still firmly attached — across the stage to the spot where the hydraulic lift lowers a section of the floor. Perhaps with a shake of the head at the antics of prima donna artists, they would merely have to start the usual hydraulics to send the embraced couple of piano and pianist down, down, into the depths of the so passionately sought out basement. There the pianist could disengage, kiss his ivoried carrier a hurried farewell, and skip down to the furnace room to see who was there that could perk up the stage-left blower for him.

This trying tale is almost exactly what the cortisol (and indeed all steroids) does when it reaches one of its target cells in the body. First it binds tight to a receptor on the cell wall, then it gets sucked down to the central depths of the cell, penetrates the nucleus, and induces the DNA control mechanism there to redirect its working and make a different blend of proteins. That changes the cell's functioning for a good while to come, and that's what makes cortisol so ideal as our longterm responder to stress.

(The whole setup can even be seen as a circuitous way of allowing your food to take care of you when you're temporarily separated from the dining table. Nibble some buttered toast with jam, and cholesterol *and* sugar will be sent down to your stomach in the very same mouthful; they only go through the rigamarole of diverging — the sugar to storage in the liver; cholesterol to cortisol creation units in the back — because that's the only way they can be stored safely against a period of big stress that might come long, long after they've been consumed.)

Of course the cortisol response is not as quick as adrenaline. Adrenaline, as we saw, works by triggering control handles on the outside of cell walls, and that's fast — as fast as if our pianist had modified the furnace by just flicking a thermostat set conveniently on the wall of the stage. But adrenaline is just a modified amino acid, and nothing like a real steroid. That means it can't get into a target cell itself, and can't set up long-lasting changes the way cortisol can. It's a trade-off: quick but transitory adrenaline; gradual but more persistent cortisol.

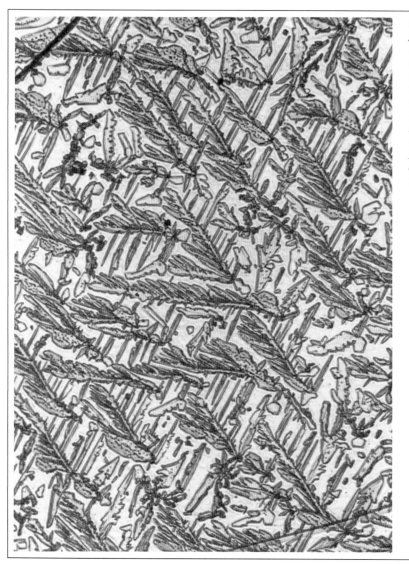

The delicate buildup of filament grains in an enlarging crystal; in human bone such even laying down in closely keyed to levels of Vitamin D — which is another derivative-form of the ubiquitous cholesterol from which our stress hormone cortisol is also made. Vitamins were originally going to be called Funkians (after their discoverer Casimir Funk) until he declined the honor.

Even the two hormones' survival in the bloodstream is different. A few seconds after starting an anger surge some of the released adrenaline is already being digested apart by your blood. Within two or three minutes all of it is gone and you can soon go back to feeling as relaxed as before. But let a constant mulling on the condo you wish you could afford trigger the wholly different channels that start the cortisol overflowing out, and then you're in for something stronger. Cortisol, being based on rugged cholesterol, can survive even amid the turbulence and acid of your blood channels for hours, there to make you hyper, uncomfortable, and possibly shaky for all that time to come. Every sudden sound will seem just a little bit louder, every move of your arms, every glance up when someone calls you, will be just a little more extreme, and more intense. It will seem as if something has come over you, and indeed something will have: the cortisol. Over a long relaxing evening the cortisol burst will wear off, and the following morning can be approached with an almost graceful ease. But let the tensions come back, let the worries bore in, and once again the cortisol-induced jitters will return. (Curiously enough, sharks and sturgeons also tremble from modified cortisol when overstressed.)

The rugged persistence of cortisol and all steroids is one reason athletes who use them right up to an event have to do some pretty unpleasant things to pass their doping test afterward. One apparently well-perfected procedure is for a steroid-taking weight lifter, for example, to empty his bladder with a powerful diuretic just after the day's competition, force a flexible catheter up his penis, and fill himself up with fresh urine from an accomplice on the other end who has not taken the persistent drug.

Athletes are only forced into this because they don't trust that their body's natural production of steroid hormones is high enough for the record-breaking performances they want. They forget that the steroids are in a delicate balance, and can only be knocked about at risk. The steroid testosterone, for example, is present at useful levels in all men. Taken in excess it will increase muscle-building (by piercing the outer walls of cells involved in that process), but it will also act as what athletes' slang recognizes as a "chemical castrater." Side effects might be a higher voice, swollen breasts, and other unwanted female characteristics. Ditto, in the other direction, for women,

Cortisol is just another part of this natural equilibrium scheme. There's a natural level of cortisol in the body, and it's only the excess amount that stress jolts on its way that causes all the problems. Even aside from stress, thirteen pounds of raw cortisol leak out from the adrenals in your back each year. Most of it starts at 3 or 4 A.M., and works neatly to perk the body up — raising blood sugars, heightening awareness — just in time for wake-up at 7 o'clock or 8. (A side effect is that it lowers your sensitivity to taste — few people care for spicy food in the morning — but that's a fair trade-off for this useful waking call.)

The main facts about cortisol were first discovered under the greatest of military secrecy. During World War II secret agents of the U.S. Office of Strategic Services reported that Nazi buyers in Argentina were accumulating vast amounts of cattle adrenal glands from slaughterhouses there to purify a cortisol hormone that would allow their long-distance Heinkel fighter pilots to stay up in the air without fatigue for hours longer than their Allied counterparts. The U.S. government took these warnings to heart, and started its own program to find a way of isolating or synthesizing these Nazi-desired hormones.

That program was given higher priority than the one to produce antimalaria drugs for the island invasions that were then going at full speed against the Japanese. After the war two conclusions became clear. One was that the OSS had gotten the story completely wrong. There never had been a Nazi plot to feed pilots adrenal cortex extract, and rumors asserting its imminence just showed the incompetence of the American spies in the field. (Since the whole program had been top secret, the exposure of the OSS's foul-up was kept top secret too. That meant little opposition when the OSS was expanded and renamed two years after the war, coming forth, with many of the same staff, as the CIA.) The second result of the rush research was more positive, for it is what led to an understanding of cortisol and related hormones years before they would have been as thoroughly known otherwise. Cortisol is not the only hormone that doubles as a part of the body's stress-control system. There's also prolactin, the chemical stimulated by suckling the breast in women that quickly gives rise to milk production, but which in stress has other effects; the natural opiumlike substances our body produces, that we looked at as pain regulators in the last chapter; and a great mix of others.

For the rest of this chapter, though, we'll switch from coping to consequences — but first it's useful to take a brief look at one phenomenon that's halfway between the two. A visit to the Wild West will introduce it nicely. . . .

Coping Interlude: Tough Guys and Tears

Federal Marshal Gary Cooper stood in the center of the deserted Western town. It was 11:59 — one minute to high noon — and any time now the mob of desperadoes were going to arrive. If this lawman was undergoing an adrenaline surge he showed it not at all; if his cortisol levels were high his cool eyes wavered not an inch. A lone tumbleweed blew ominously past the shuttered, silent homes.

The camera moved in for a closeup. Slowly, barely perceptibly, Cooper's rigid jaw began to move. His lower lip started to tremble, his cheeks started to go red, and curious spasms of muffled sobs came over him. Tears welled from the unfortunate's suddenly puffy eyes; strange noises: "mn-ah-oooh-bhoo-h" came from the now-contorted lips. Great Cooper was in tears.

OK, so that one ended on the cutting-room floor. But why is crying so frowned upon, so derided as the most opposite extreme from the toughened, manly ideal? The mechanism certainly is straightforward, if only slightly disgusting. Tears are modified blood that has had its few coloring cells removed. As we referred to in chapter three, tears are squirted out like a car's windshield fluid — eight pints a year — then held tight on the eyes, sandwiched between a layer of mucus on the bottom and a layer of oil on the top.

Normally tears drain into little holes on the side of your nose (when blinking slows down, as at night, they dry there and form what we call sleeping dust), and only overflow when so much is poured out that they burst through the tarpaulinlike web of oil overhead. Blinking of the eyelids (each blink: $\frac{1}{5}$ second; total blinks in a lifetime: five solid years) creates a roaring low-pressure wave that helps to send the overflow back into the holes to the nose; indeed, over-fast eyelid fluttering is usually the last giveaway of a crying fit about to break.

Why cry? Is it that dangerous stress hormones are conveniently ejected with the tears? If true, that might explain why women, who are allowed to show their tears, have generally better health than men ($\frac{1}{4}$ the ulcers, and a four-year-greater life expectancy); also why executive women, who have less chance of crying, do not live longer (only a one year greater life expectancy than men for them).

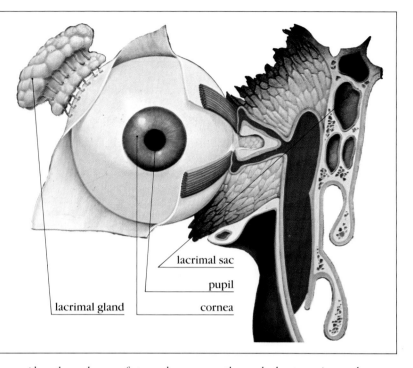

Tear flow across the eye. Production is from the gland above and slightly left of center; exit is through the inner corner of the eye into lacrymal sac down by edge of nose.

lacrimal sac

pupil

lacrimal gland

cornea

Alas, the release of stress hormones through the tears is nowhere near abundant enough to do this. It's more likely that tears are important just for the signaling of personal distress that they give, and the bonds of social compassion and aid that they call for. By this reasoning women are allowed to cry because it's acceptable for them to ask for support from those around them. Men by contrast are not allowed this secretory call for help: not in a society like ours, where they are supposed to show that they can stand on their own — and where the ideal marshal has to face the desperadoes from High Noon central casting alone.

Stress Consequence: The Vagus

Some stress responses are less ambiguous, and only give everyone grief. Take the nerve that is possibly the weirdest one in your body: the vagus nerve. The name comes from the same root as *vagrant, vagary,* and *vagabond,* and that image of directionless ambling describes exactly what it does. The vagus nerve wanders in the most extraordinarily aimless fashion. It begins heading down from your brain, funneling through a special hole drilled for it in the skull. Then it sends out shoots in all directions. Some reach to the lungs, some curl into the throat, and other traipse all the way into the center of the heart. The vagus gets in them all.

The vagus's branch in the heart will show just how important this nerve is. Normally the vagus branch in the heart has the job of keeping the heartbeat at a reasonable speed, calming it down from the inbuilt and far too dangerous 140 or so beats a minute it's straining for all the time. But when you send the contents of your first wok dinner flying with one too forceful spin, or when you worry for the tenth time about the imminence of the IRS deadline, and the incompetence of your chosen accountant, then the vagus nerve section in the heart gradually slows down its firing. That cuts down its restraining effect on the heart, and the result is that the heartbeat leaps up to a racing eighty or ninety pulses per minute, where it will stay until the vagus nerve comes back into full effect and the body has finally calmed down.

Other parts of the vagus nerve are just as strongly affected by stress. Some, the ones that produce faster movement or tingling ears, are easy to pick up. Other effects don't make themselves known for a while, but since they all depend on the vagus nerve, are happening just as much. Of these perhaps the most unbecoming is what a worry-stricken vagus nerve does to your digestive tract.

First it clamps down on the little-remarked sphincter of Oddi. This is a little tunnel, discovered by an inquisitive Mr. Oddi, that carries digestive bile juices from the gallbladder into the top of your small intestine. Normally the digestive bile comes in at regular squirts, but if you get tense for long enough the vagus nerve branch there will keep the odd Oddi tunnel closed so none of the digestive juice can get in. The food you might have in there sits and waits, most uncomfortably, until the vagus goes back to normal and lets things proceed.

That's bad but there's worse. Another strand of the ever-stretching vagus nerve wends its way to the middle part of the large intestine, almost at the very end of the whole digestive tract. What it's supposed to do there is give the already digested food that's made it that far a good walloping push over to what is the end of the digestive tract, the descending colon, so it can be expelled from the body when the mood and opportunity arise. That's what it's supposed to do. But when stricken still by worry, it does no such thing. There's no vagus message to the middle of the large intestine, no satisfactory wallop over to the final section, and so no expulsion of the digested food at the time that is best for it. Instead, the material sits there not quite all the way along, waiting for a signal that does not come.

That gives the rapacious large intestine its chance. It sometimes seems that the large intestine's main goal in life is to extract as much water as it can from whatever it is that gets in its clutches, and now, with some overdue digested food there, it really gets to work. Hundreds of thousands of minute octopuslike tendrils reach down from the walls of the large intestine, and greedily slurp up all the water they can get from the waiting digested food. A few extra hours of this doesn't matter much, but if it goes on for two or three days, if the body's owner is worrying or ill at ease so much during that period that his vagus nerve stays numbed into inactivity, then the water extraction does begin to matter.

The digested food residue that eventually will have to be expelled gets drier, and harder, and tighter. You might almost say the food pieces get crammed together, and that is exactly what the mighty Imperial Romans, suffering on their stone-carved toilet chairs, did say: they had become *con-stipati* they shouted out, meaning "crammed together" — a phrase that comes out in English as "constipated."

The large intestine, all six feet of it. Flow is upward on the left, across on top, and down, to the anus and final expulsion, on the right. Feces (pink) are more diffuse on left and top than on right, due to shorter time in the large intestine and so less water absorption. Note the appendix waving bottom left, hip bones and spine as green in background.

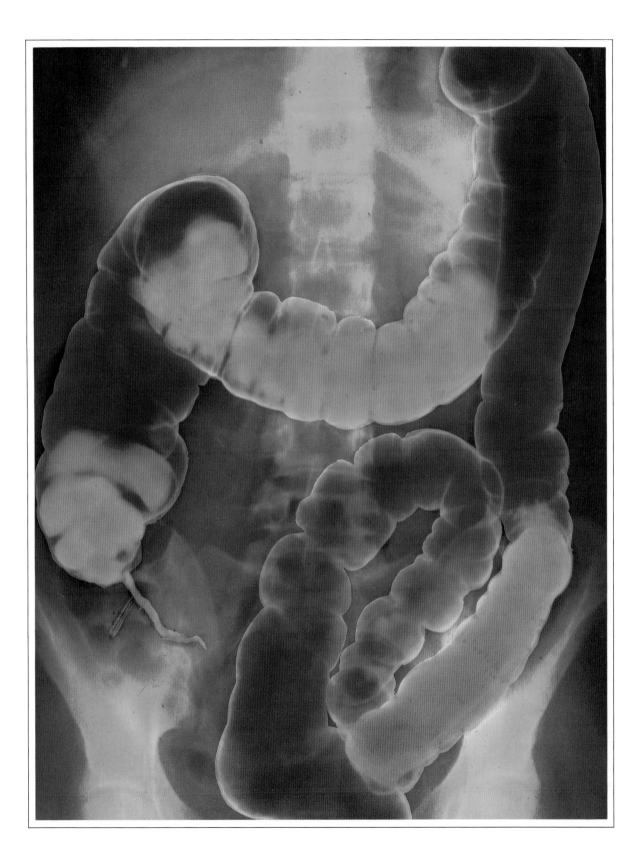

From this overlong exposure of digested food to the water-grab-bing clutches of the large intestine, comes the pushing, the straining (which can raise internal abdominal pressure to almost 100 pounds per square inch, over seven times atmospheric pressure), and the painfully mixed desire and apprehension about entering the bathroom for another scene of battle. All from the ever-wandering vagus nerve, and its insidious inaction whenever the stress builds up.

Stress Consequence: The Stomach

Another specific body change that happens when you worry and get anxious too much is the stomach ulcer. It afflicts perhaps only one in ten of the population, but when it does come, its sharp pain, occcasional bleeding and general bother more that makes up for the spars-ity. But why the stomach? Is it all the excess acid, the stuff advertisers warn us about when hawking their antacids and bicarbs, their Alka-Seltzers and Milks of Magnesia? Not quite.

The stomach is an extendable bag that developed about 500 million years ago when certain early fish needed someplace to store the chunks of food they could so quickly bolt down with a new piece of equipment called the jaw. (Before that no animals had any jaws, and were never likely to munch more from their prey than they could di-gest at the moment.) In man the stomach can hold about three pints, in dogs about five pints, in wolves, which wolf their food, over eight pints, and in cows, with their four-chambered stomach and a need to hold vast quantities of grass long enough to ferment, the stomach can hold forty gallons — as much as a good-sized bathtub. Those who have seen a cow regurgitate the full extent of this awesome holding report that it is a sight never to be forgotten, and seemingly never to end either.

In all animals the stomach will freeze when cold, going numb and motionless for half an hour after eating ice cream. It will also glow bright red after touching pepper, be it the choicest French *steak au poivre,* or the meanest roadside Tex-Mex chili. When it's empty and wants more, the stomach will squeeze its walls, compress the few ounces of gas always inside, and send the sounds of that compressing gas through the megaphonelike membrane of the abdomen out into the world. This is the origin of the often conversation-stoppingly loud gurgles of a belly rumbling strong. (The technical name for these sound effects, delightfully, is borborygmus, Greek for "stomach rum-bling.") While definitely the cause of that auditory affirmation, the stomach is in fact located much higher than most of us think, way up near the heart, and so is unjustly blamed for the various squelches and gurgles that emanate from the small intestine lower down.

To see how stomach ulcers come from stress, it's best to start with what happens after a normal meal. Inside the stomach wall are dozens of little wrinkles that only straighten out when the stomach is stretched by a good meal. When that happens a churning process be-gins, languorous and in slow motion, that can take half an hour to reach down to the center of whatever it is that has been swallowed. Since a few colonies of bacteria get swallowed with every mouthful

Where ulcers start: the inside of the gallbladder. The small circles are individual cells, which pump out electrically charged particles to help concentrate the digestive bile that is sent over them by the liver. Foldings are the body's natural attempt to increase the number of such cells. Result: bile salts powerful enough to digest the stomach wall. Normally interlacing muscle sheets just under the surface tighten the gallbladder and send the load into the small intestine as soon as it is ready, but stress-induced backflow there can allow it to reach the stomach and attack. Note: green is broken-down red cell constituents, which, when acted upon by bacteria in the intestine and expelled, will give feces their distinctive odor; yellow is what Greeks called solid (sterol) bile (chole) — our cholesterol.

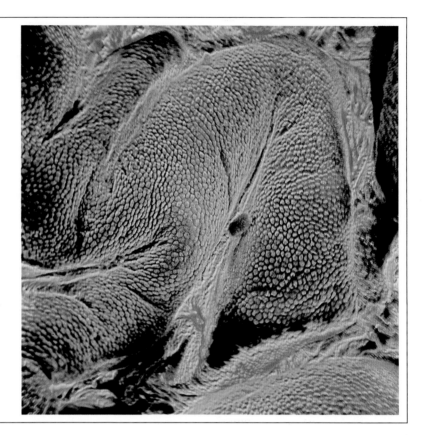

you take, the curious spectacle arises of a whole generation of new bacteria being born, reproducing, a dying of old age, before they are reached in the food ball at the center of the stomach.

All over the stomach's wall are tiny pits, 35 million in fact, and from even tinier pipes opening up in each one a steady rush of gastric juice comes out whenever food is on its way, or the eater merely hopes it is. That juice is 0.5 percent pure hydrochloric acid, the stuff that medieval alchemists came up with when they needed something to dissolve containers of heavy cast lead. The hydrochloric in the stomach is just as strong as that, and is perfectly capable of interacting with the digestive enzyme pepsin to dissolve a thick steak right away. Boa constrictor snakes share the same acid with humans, and as their diet of full-sized living donkeys and pigs shows, it's perfectly capable of dissolving away living tissue along with supermarket offerings too. Hydrochloric is so potent that your body keeps it carefully separated into innocuous hydrogen and chloride fragments until it's actually needed. Then, like a binary chemical weapon fusing the two fragments in its warhead, two pumps force the constituents out of the cells where they're stored and up along the tubes to the stomach.

Now if pepsin and hydrochloric are so corrosive, the worrying question is how come they don't digest the walls of the stomach itself while they're at it. The reason they don't — usually — is that all over the inside surface of the stomach is a single layer of cells that look like

a field of baseball bats balancing straight up on their handles. In the stomach their ends are so tightly wedged together that none of the acid that made it into the stomach to digest the food can get back through them to start digesting away the wall of the stomach itself. Some 800 million bat-shaped cells are wedged tightly in that defensive position all the time, and if some get worn away, others are likely to spread out and fill the breach. In fact, half a million fresh wall-defending cells pop into existence in your stomach each minute, enough to create a new liner, pink and shiny fresh, every three days.

With all that going for you it seems amazing that a hole in the wall

An ulcer: wearing down through the protective layers lining the inside of the stomach to the delicate underlying tissues themselves.

is ever kept open, and a stomach ulcer ever produced. But, oh, it does happen, and the initial culprit is not excess acid, not hydrochloric nor sulfuric, but instead a sloshing backflow of bile salts from the small intestine where it leads out normally away from the stomach. These bile salts are formed in the liver from the ever-ubiquitous cholesterol, but in a wholly different way from the stress hormones. What the bile is supposed to do in the intestine is shake up big fat globs into tiny ones that can be more easily digested. And in the intestine that's fine. But when the normally tight sphincter barrier between the stomach and intestine is loosened by persistent, heavy stress, the bile salts can shake their way into the stomach. (Our friend the vagus is largely behind this, loosening that barrier, and even raising hydrochloric output to boot.) Then it's all downhill. The salts shake apart any fat globules they find there, which is not so bad, but they also jostle their way through the defensive bat-shaped cells lining the sensitive inner wall of the stomach, which is.

Through that single breach the acid in the stomach can pour: great tissue-frying juicy streams of it. Combined with the irritant histamine released by nearby cells in there, that burns millions of wall cells away. Happening once or twice, the stomach wall's natural rebuilding ability will replace the damage within half a week. But kept up day after day, at the office and after work, the hole will widen and enlarge, hurting as it goes, making every worry an all too precise flash of heating pain. Kept up hard enough it might even hemorrhage, and that is call for immediate surgery.

Not all stresses make ulcers. Only the right kinds do. It used to be a saying about the steel-hearted accountants on GM's fourteenth-floor headquarters, that "they don't get ulcers, they give them." More scientifically, if not notably more kindly, researchers have found that monkeys who had to press a bar to fend off electric shocks that came at them whatever they did, developed ulcers at a near 100 percent rate. If the shocks came at them on a regular schedule, they didn't. The person who develops ulcers is not so much the one who has struggled to get ahead, as the one who is struggling still.

Stress Consequence: The Head

Headaches may strike in the head, but they actually begin down in the incredible factories that make up the marrow of your bones. For that is where your platelets are formed: sticky objects, each smaller than a blood cell, that pop into existence when great production cells in the marrow squeeze out a tentacle arm, then bubble off the platelets as little pimply dots. Hundreds of millions of platelets get born in your bones this way every day, and wait there only a moment, oblivious to the crawling scavengers that are burning away calcium struts in the bone architecture beside them (bone is an ideal storage depot for toxic minerals such as calcium), before rising up, floating loose, and starting on their voyage through all the reaches of the bloodstream.

Each platelet is crammed full of a tremendous variety of supplies: There are cords and tubes and energy batteries and bacteria killers and clotting agents and even small baskets of adrenaline. Some of this is useful in helping blood clots form when there's a dangerous cut or a burn, of course, but in headaches what counts is that the platelets also have in them yet other baskets, these contain both the acidic substance serotonin, as well as precursors for a nasty hormone that's actually a part of cobra venom: kinin.

A long bout of stress — worrying about the mortgage payments; putting up with preschool kids underfoot, or just a spouse who persists in acting like one — can make platelets that have stumbled into the blood vessels around the scalp start to adhere together in great gooey clumps. When these clumps get found the platelets split and their baskets containing the serotonin and other substances come trickling out. As the serotonin emerges, it briefly forces the muscles wrapping around the small blood vessels in the region to tighten and constrict in a band, so raising blood pressure on the unsuspecting scalp, and also constricting the flow. After that misdeed the serotonin streams in toward the brain, lowers the normally protective barrier the brain keeps up against intruders from the body, and coolly lets itself in. There the serotonin speeds message transmission in the electrical circuits that make up the centers regulating our emotions; the jolt of its arrival could be what causes the sudden bad temper headache sufferers so frequently lash out with.

For the throbbing pain of headache we have to turn back to the outside of the head, as the brain itself has no pain endings, not a one (so much so that brain operations can be carried out with the patient fully conscious, and only local anaesthetic used to numb the scalp when the first few cuts are made). The kinin hormone leaked out by the colliding platelets is one of the most potent blood vessel wideners known (as we'll see in chapter seven), and here it undoes what the serotonin started and increases blood flow to the scalp with a vengeance. the kinin also cruelly sensitizes the pain receptors alongside the blood vessels throughout the scalp and neck, and more pain there is just what a headache means.

At the same time, yet another undesirable substance is being secreted in the blood vessels around the head. This is a fatty acid that has the charming habit of making smooth muscles go into spasms and recoil wherever it touches. A distant relation of industrial detergent, this substance seems to be behind the pain of menstrual cramps as well, and here does its bit to make the tenderness and ache of headache more painful still.

Is there any defense? Try not scowling or frowning. Those actions constrict blood vessels in the face and neck, and that, apparently, is a sure-fire way of making your platelets crunch together and release the serotonin and kinins you really wouldn't wish to be there. Not all your body's platelets will burst, of course — not even a decent fraction of one percent: only enough to make a headache throb in pain.

Insincere smiling will call out the unwanted platelets too. Over seven miles of tiny blood vessels are contorted and compressed when you hold a smile for just a five-minute stroll the length of a crowded cocktail party or business social. Some seventeen main nerve endings on each side of the face, and thousands of individual nerve contact

Two exhausted rhinos. Their overall metabolic rate is regulated by the potentially stress-affected thyroid gland, much as ours is — another sign of the zoological links in our hormone system and the stress response. It was in this species that the parathyroid glands — located near the thyroid, and involved with providing the right balance of chemicals for muscle and nerve firing — were first discovered.

points spreading direct from the brain, are called into action to hold the muscle segments needed to lock the smile in place. Indeed, the unpleasant masquerade behind such contortions is suggested by the origin of our word "grimace": it resembles the Old English *grima,* meaning an actor's mask. Truth in advertising will here do wonders.

Grimacing leaves its mark, at least for those over forty. It's been estimated that 200,000 frowns will produce one permanent wrinkle, and although the precision of that figure is possibly in question, the direction it points out is only too true. All the substances in your skin that normally keep it smooth — the flexible sheets of collagen, the rubbery braids of keratin — become terribly fragile with age. Each time the face is stretched in an excess smile or a frown, some of the collagen sheets in it will break forever; some of the keratin coils will crumple and never properly refold. Combined with the drying and loosening all aging skin has to go through, that adds up.

Broad Consequence of Stress: The Thyroid

Long-term stress gets to you in even less apparent ways. There's a certain peaceable object in your throat, that normally gurgles contentedly and just works to help keep the body's metabolism in order, but in stress will suffer with all the rest. The effect here is a minor one, but what's amazing is that the stress from job or home can work its way into this putterer on the throat at all.

The thyroid gland is a pinkish, butterfly-shaped organ, which wraps around the front of your windpipe at a point just below the Adam's apple. It gets its name from the niche you can feel on the point of the Adam's apple, a niche like the ones the ancient Greeks always carved on the top of their war shields. *Thyreos* it was called then, whence our name "thyroid" for its less fearsome imitation now.

The real work of the thyroid gland goes on in the center of its two evenly splayed wings. While these just look smooth and pink to the eye, and gooey and a bit more pink under a magnifying glass, in fact they are made up of millions of tiny little jelly bags, each one surrounded by a thin yet toughened layer of cells, and further surrounded by a close network of tiny blood vessels. Inside the little jelly bags is where all the important work goes on. That's where the hormone called thyroxine is made.

Someone born without the ability to make thyroxine would never grow bigger than a seven-year-old, never mature sexually, and never be able to speak or hear. The deficit is so clearly not the sufferer's fault that even simple people are likely to treat the victim with at least some charity. In fact, Alpine farm laborers in the Middle Ages who noticed this disease felt they saw a sign of God in its victims, and called them, quite tenderly, Christians. This came out *chrestien* in their language, so it's "cretin" in ours.

Luckily it's a rare affliction. The little jelly sacs at the center of our thyroid glands are usually only too eager to go. How they work is to pluck in individual molecules of iodine from the blood vessels outside the sacs, then pluck in special amino acids from the blood too, and whip the two together into whopping great globs of thyroxine, so big that they have no chance of slipping out of the sacs before they're called for. It's easy enough to get the amino acid that goes into making the hormone, for the body can cook that up from almost any food that you eat.

Getting the iodine can be more of a problem. Not a lot is needed, really just 1/150,000 of an ounce a day, but that little cannot be missed. The way it gets to your thyroid is amazing. The route begins, of all places, at the point where raindrops from massive storms plunk down on the ocean. A single raindrop landing in the sea doesn't make much of a splash, but it does make enough of one to sling up into the air a bit of water that most likely has a sliver of iodine in it. In the time it takes the splash to lift and fall, some of the water will have evaporated, and so will the odd iodine bits that had been floating along within it.

On these evaporated bits has the continued survival of mankind depended. Most of the iodine molecules fall down again into the sea nearby, but a lucky few get blown all the way to shore. As might be expected, the greatest number flutter down in regions bordering the sea. Because of this, fruit grown in California or New Jersey naturally has a lot of iodine; fruit grown in the Midwest does not. The concentration of iodine in your thyroid is 60,000 times greater than it is in the rest of your body, and in the thyroxine molecules (each one has four iodine atoms) it has to be 2,000 times higher than that. Without the iodine supplements the government sees is put in their table salt, residents of

Close-up of living thyroid tissue, where the immensely powerful hormone thyroxine is being formed.

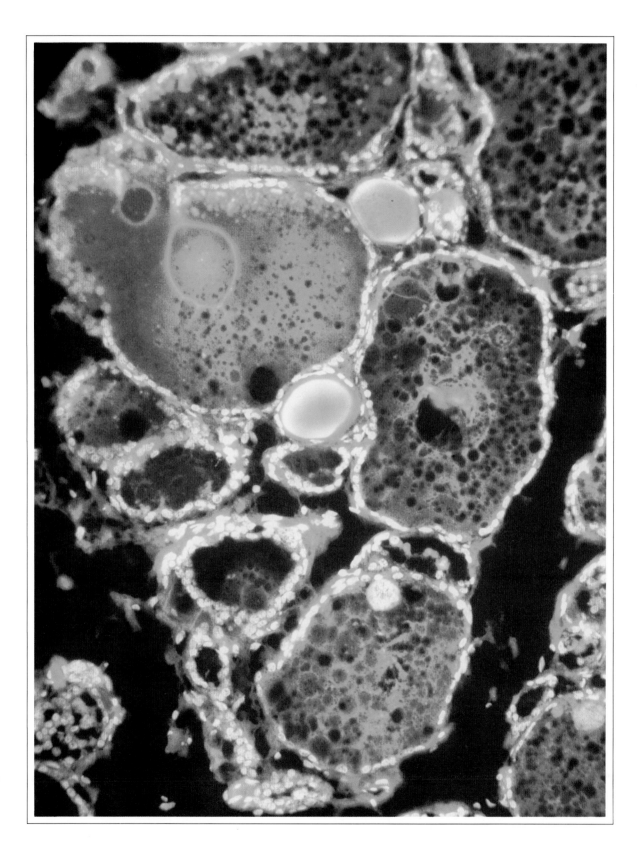

Chicago, Kansas, and similar central outposts would find their throats swelling up as their thyroids built on extra jelly sacs, lots of them, to pull in what little iodine might be available. (This condition, now rare because of the iodine supplements, goes by the name of goiter. In the 1800s it created what was called a "goiter belt" all through the newly settled, and iodine-poor, Midwest.) Normally, though, the right number of thyroxine chunks are made at the center of these little sacs, and normally the right daily does is triggered to go off into the body: 1/110,000 of an ounce, no more, and no less.

Once slit from the gobs they're attached to, the thyroxine gets swooped out of its jelly sacs in the neck by the pull of the blood vessels passing all around. Then it's off for a ride in the bloodstream, becoming stitched to a new blood protein, and converted into a form with one less iodine atom, as it goes. Wherever the fresh hormone arrives, it triggers cells into churning at a higher rate, to produce more heat, burn more oxygen, and break down more glucose. That's what one thyroid hormone molecule starts in one cell, and in the daily dose of 1/110,000 of an ounce, there are quite a few more than one thyroxine molecule. There are fleets of trillions. A lot of the body's cells get these amphetaminelike speedups, and the result is that tissue growth and development is promoted almost everywhere. That's indispensable for keeping you fickle, active, and fit.

This is all quite fine, but so far we haven't seen what it has to do with stress. Thyroxine is usually triggered to be let out of its production sacs at just the right rate to keep the body's metabolism going at the level that's best for it. The normal amount keeps you eating 2,500 calories a day, say, to keep going. A single extra gram would supercharge you so that you would have to gobble 3,500 calories just to stay even. But since thyroxine is an immensely powerful hormone, any change in the rate at which it is converted into the proper three-iodine form will slow the whole body. And that change is unfortunately only too easy to produce with the aid of a good bout of stress. Start thinking about an important transatlantic phone call that's supposed to come, or dwell at length on the ideal apartment you want that's just one bit out of your price range, let all this happen often enough and gradually some of the three-iodine form of the hormone will be switched yet again into a shape that doesn't work as well at all. That drags down your metabolism: some of the cells that were stimulated before will miss their full, daily hit. It's a subtle effect, and it takes hours to show itself, but it does serve to show just how many convoluted but crucial pathways there are inside us, and how daily stress has the ornery power of getting into them all.

As a minor consolation, other species can be affected by thyroid hormones even more powerfully than man. Graduate students in hormone studies regularly get a chance to see their power by an experiment on tadpoles, those fishlike juvenile forms of frogs that can give even the meekest of bookish students delusions of being a god. Certain tadpoles will spend two years swimming contentedly around, steadily growing, until they get ready to make the switch into frogs. These are the kind that turn into bullfrogs. What the graduate students are told to do is take one of the young tadpoles out of its pond, one that still has almost the full two years of growing to do before it switches into a frog, and feed it a single drop of pure thyroid hormone. When that happens: BOING! The tadpole bursts out, there and then, into a perfectly formed bullfrog. Legs shoot forth, the tail gets sucked in and disappears, and a gnarled face and bulging eyes soon appear.

What's most striking is that all this happens without the tadpole lengthening by a centimeter. It turns into a fully formed bullfrog, but it does so at its original size. The result is a bullfrog half the size of the nail on the little finger, puffing and croaking in all its glory; a bullfrog that has no idea that in two years time it is going to be dwarfed by its once brother tadpoles. When they develop it will be into full-sized bullfrogs, 80 or 100 times bigger than the thyroid-advanced little one. There is likely to be one last little croak, a hasty gobble from one of the giant bullfrogs, and a hasty end to the subject of this experiment in development superacceleration.

Broad Consequence of Stress: The Immune System

An even more clear-cut way the consequences of stress will work through your entire body is what happens to the immune system. Your immune system can be considered as a single organ that just happens to be spread in bits and pieces all across your frame. In this perspective your bloodstream is a single dispersed tissue too, weighing about twelve pounds, while the skin is yet another, weighing perhaps over fifteen pounds. The immune system weighs just two pounds, but contains in that over a thousand billion defensive white cells, and — watch this now, typesetters — over 100,000,000,000,000,000,000 attacker antibodies. Give or take the odd 10,000,000,000,000,000,000, that is.

These scattered defenders continually die and have to be created afresh, which is why when you worry and stress slows their production you get sick: more colds, more flus, more everything. It's not just something you imagine — it really happens. One major Australian study, for example, has shown that widows and widowers have a lowered immune response for several months after their spouses have died, and that they're more likely to get a host of infections — possibly even cancer.

One mechanism starts right in the middle of your chest, just at the point you hit when you tap your chest. That's the home of the matchbook-sized, pinkish gray glob known as your thymus gland. It's a vital part of the immune system — building up killer white cells; controlling antibody regulators too — and without it we would be in pretty bad shape. Take a child's thymus out surgically and he could literally start getting eaten away by the ever-hungry "bugs" that are always present in the air and inside us. Fungi might start munching away the fingernails, streptococci could clamber onto the delicate heart valves and chew them to shreds, and the inside of the mouth might start rotting away under algaelike growths.

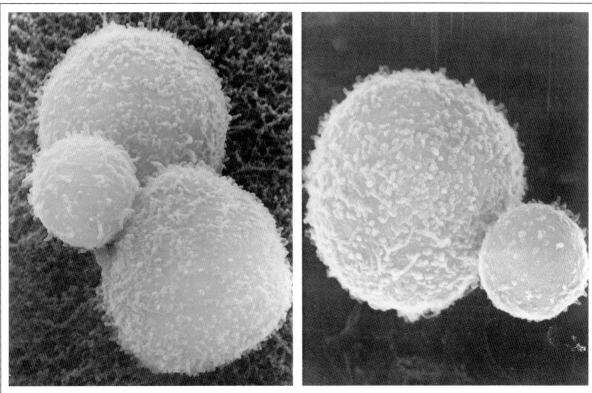

Battle in the bloodstream — killer defensive lymphocyte being warded off by a cancer cell. This series shows the dangers of not having enough defensive white cells, as happens when long-term stress keeps up.

Small sphere in first micrograph is one of body's defensive white cells, called a killer lymphocyte. It was produced in the thymus gland, and has here come into the region of two cancer cells (larger spheres). In the second picture killer lymphocyte recognizes surface structures (antigens) on cancer cell, and closes exclusively with it, drawing it aside. In next picture the killer lymphocyte is mounting its attack, causing the cancer cell to lose its small surface projections, and blister and begin to bubble away.

Normally other lymphocytes would be available to join in and ensure that the cancer cell was destroyed. But under stress, their source, the thymus gland, can shrivel to less than half its usual size — meaning the killer lymphocyte might have to face the cancer cell alone. In the fourth frame we see what happens when it's only one to one: bubbles coming out from the cancer cell form into neat spheres and head straight for the lymphocyte. In the final frame they form a barrier phalanx between the now-recovered cancer cell and the lone, unsuccessful "killer" lymphocyte. Cancer cell survives, perhaps for future growth, due to stress-lowered immunity levels.

Swelling in neck, armpit or groin is often a sign of such encounters, for that's where lymph nodes — where lymphocytes gather — are concentrated.

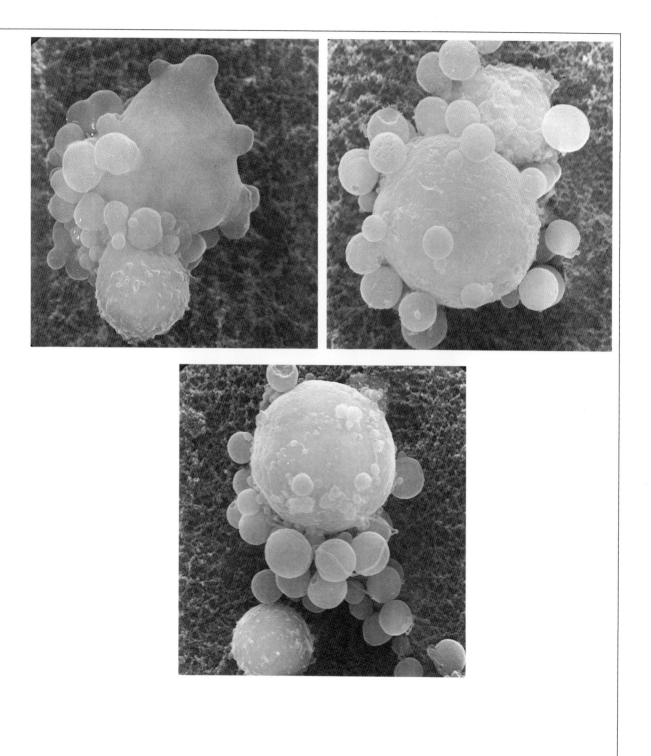

Worry — the scholar's indispensable lot: captured perfectly in this gargoyle from the bell tower of New College, Oxford.

The effects after you've agonized through a long period of stress won't be quite so bad, but the cortisol steroid that automatically comes out in stress can go a good way toward slowing the immune system enough for you to suffer. Cortisol sets upon the thymus and shrinks it; tracks down helpful white cells that have escaped and dissolves them. (Cortisol is so strong this way that it's given after transplant operations to help keep the immune system too weak to reject the new graft.) The result is more sniffles, sneezes, and fatigue; possibly more serious debilitation, too, as intruding germs take advantage of the stress-weakened you. Why does your cortisol do this to you? It might just be sheer bad design. There are certainly enough uncoordinated muddles in us to back this: folic acid, for example, the stuff we need to hold back anemia, is continually being churned out in the large intestine, but is no good for us there as it can only be absorbed by the small intestine that its constituents passed through earlier on. More likely, though, cortisol is used as a useful regulator of the immune system — slowing down the movement of white cells from the thymus; slowing down their growth when needed, too — that only in stress gets pushed overboard to make us suffer.

To end, a note of caution. Or rather against caution. This chapter has treated stress as a calamity to be avoided, but life entirely without it would be horrendous to behold. Imagine all your family grinning docilely around the breakfast table in the mornings; taxi drivers stopping to let stray caterpillars safely cross the road; your colleagues reading inspiring passages from the *Reader's Digest* to each other at work when you arrive. You would think someone had been tinkering with the city's water supply; after a few more hours you'd probably be desperate to head over and tinker it right back.

Stress in controlled amounts has plenty of worthwhile uses. There would, for example, be no sensation of overcoming obstacles without it. As one delightful example, Dr. Truman Stafford, Harvard instructor at age twenty, was once asked to square an eighteen-digit number in his head. According to an onlooker, he then "flew around the room like a top, pulled his pantaloons over the top of his boots, bit his hand, rolled his eyes in their sockets, sometimes smiling and talking and then, seeming to be in an agony, blurted out an answer: 133,491,850,280,566,925,016,658,299,946,583,225." That was the correct solution to the problem, and young Truman was overjoyed by the result. Spared the agony, he would also have been spared the ensuing pleasure.

Another example of the occasional indispensability of stress is that of a certain twenty-eight-year-old Austrian immigrant working at the University of Montreal in 1936. He was trying to purify a substance, in the hope of finding a new sex hormone in it, when suddenly he realized that his procedure had been wrong and that there was no chance of a new hormone being found. As he put it several decades later, "I do not think I have ever been more profoundly disappointed! Suddenly all my dreams of discovering a new hormone were shattered."

At first he did what would be considered the sensible thing: "I tried to tell myself, 'You must not let this sort of thing get you down. . . .' I tried to tell myself over and over again. . . ." It didn't work: "But all this gave me little solace. . . . I just sat in my laboratory, brooding about how this misadventure might have been avoided. . . . Somehow I could not get hold of myself sufficiently to do anything else in the laboratory for several days."

These few days of brooding stress were crucial. Suddenly the young Austrian saw that while he was not on the right trail for a new sex hormone, he did seem to be close to discovering a whole range of stress hormones and their controlling factors. He became entranced by these unexpected possibilities he had come up with in his despondent dwelling on the problem: "Instead of dropping the . . . problem . . . I was now prepared to spend the rest of my life studying it. I have never had any reason to regret this decision."

Hans Selye was the immigrant's name, and when he became one of the world's most respected biologists for his stress discoveries, discoveries that are at the core of current scientific work on stress (as well as much of this chapter), he didn't have any reason to regret his decision either.

Conclusion? Stress on the whole is bad, but a certain amount of it can be indispensable. The question is how to choose the balance. A motto Selye came up with gives perhaps the answer we would do best to hold on to:

*Fight always for the highest attainable aim
But never put up resistance in vain.*

He knew.

7

Hot and Cold

*I*n the early 1600s, an Italian doctor with the euphonic name of Santorio Santorio built a vast weighing machine big enough to hold him, a table, a chair, a silver chamberpot, and a few of his favorite books. He sat on the balancing platform for weeks on end, while his servants kept four bonfires going around him. Everything was carried out with the utmost seriousness, for Santorio was convinced that he was about to measure the weight of his soul.

At the time most professional investigators believed that the body's heat was generated by a divine spark in the human soul. Santorio, up on his platform, was one of the professionals. Although his experiment showed only that a man surrounded by bonfires will sweat away a lot of weight, no one thought him odd for trying it. Only the Wars of Religion later in the century, which saw the massacre of many of Germany's inhabitants without any change in the proportion of Catholics to Protestants, was there a gradual move to the belief that perhaps human beings were not endowed with a fragment of the divine spark after all. Something had to be found to fill the gap, and that something was food.

Heat Production

Although little bits of celery and rabbit thighs were rather less impressive than a spark of the divine presence, they did have the advantage of fitting in nicely with the newly developing scientific view of the world. And in that view food had the great advantage that it could be seen, measured, quantified, diluted and, above all, it could be burned. Not much, but still enough to see. Hold a cashew nut over a fire and it shoots into flames. Hold a piece of meat over an intense enough flame, and it too will ignite. The reason is that the strands of protein that once formed a steer's muscle, or the delicate sheets of material that once formed the nut of a growing plant, are held together extremely tightly. If they had just been resting against each other, the steer would have collapsed in pieces every time it moved and the plant would have flown apart every time the air blew over it. No, the component parts of food hold together through chemical latches that are hard, very hard, to undo.

To make them come apart, something very strong has to slide up onto the latches, something like a powerful enzyme perhaps, and give them a good hard rip. This makes the parts snap apart at high speed, and this quick unsnapping is precisely what we call heat. A pound of butter unsnapped in this way will not produce as much energy as a pound of TNT, but enough so that Arctic dog-teams can be kept warm in −60 degrees weather by giving them sticks of butter to munch.

This is what happens in the body over and over again. The chunks of food that make it into the stomach are a lot bigger than the body can usefully handle. The only way to cope with such gobs of food, so enormous to the individual cells, is to break apart the chemical bonds that hold them together. And doing that releases heat: a bit in the stomach, a lot in the intestine, and even more in the liver. Even then, there's worse to come, for the little pieces of broken-down food that float out of the liver into the arteries make it to the leg muscles, heart muscles, earlobes and eyelashes. There they are crunched apart by other enzymes when the time comes, giving off more heat in the process.

This heat production goes on all the time. Hitting an elevator button produces 1.6 calories, getting shaved or brushing the hair 12 calories, making beds 30 calories, and walking 200 calories in an hour. These are averages: an Olympic butterfly swimmer can heat at the rate of 650 calories an hour; someone gently doing sidestroke lengths to unwind in the evening might produce only 250. Still, an average person's diet releases in one day as much heat as the burning of a pound of coal into ashes. This goes on day after day, so somehow the heat has to go away. If it didn't we'd melt: an intrinsic heat production is enough to boil away a bathtub full of ice water in less than a week.

As is so often the case with the body, the bloodstream comes to the rescue. Heat is easily absorbed by water, and the bloodstream, which is 80 percent water, is continually passing through the hot regions in the center of the body, and weaving out to all the points along its surface. There the heat is close enough to the generally cooler air around us to be wafted away. It takes only 40 seconds or so for the blood by the steamy hot liver to flow to a slender vessel that might be tucked less than 1/20 inch from the surrounding air.

About 70 percent of the body's excess heat gets carried away in this way. Another 15 percent of the heat that piles up inside us when food is broken down is removed by sweating, 7 percent by evaporation from the lungs, 3 percent by heating up the air that is sucked into and out of the lungs, and the remainder in the urine and feces. That's if everything is working quietly under normal conditions. Sometimes it's not, and that's when the really important backup systems come into play, for a creature that could give off the heat it accumulates only when resting quietly would face a dilemma when a hungry predator came into the vicinity. If it stayed still and cool the predator would get it, but if it tried to run for it, it would generate excess heat, a lot of excess heat, when the little bits of food resting near its muscles were split apart to provide the energy for running away. A teenage girl going out for some hard track exercise can build up in four hours as much heat as she usually generates in four days, and a middle-aged pudgy executive would not do much worse. If special mechanisms were not efficiently built in to get rid of the heat, such activities would quickly be fatal.

It's just one of the many limiting ranges that the body is carefully rigged to stick to. The arteries can't withstand blood pressure above 400/200 or work at blood pressure below 60/30; total electrical levels in the brain must not vary much from 20 watts; inhaled carbon dioxide must not exceed 6 percent, or atmospheric pressure fall below a quarter of sea level and so on for every bodily system. When subjected to a setting away from its usual equilibrium in the middle of a range, the body almost always begins to right things. These efforts can all be considered examples of the stress response (see chapter six) and temperature control, to keep the body in the narrow range around 98.6 where it works best, is a prime case in point.

Heat Dispersal: Sweat

The backup system that has evolved for getting rid of such sudden heat loads, so allowing primitive creatures to run away and primitive predators to lunge after them, is the much-maligned action of sweating. Because the word *sweat* comes from the relatively understandable Middle English, pedants and schoolteachers for centuries, have preferred the more obscure word *perspiration,* which comes from the Latin for "breathe through." It was used in medieval France, so acquiring the desirable pedigree, but it was used because people then mistakenly believed that the pores sending out sweat also acted as mini-lungs and blew air out. Both names are actually misnomers, for what people think of as sweating takes place only when real sweating hasn't worked. The popular image of sweat is of a disreputable clamminess, an embarrassing sticking of the clothes and a wetness in various dark places. But this sensation is just an awareness of water on the surface of the body. The interesting matter is what it's doing there in the first place.

Whenever the brain decides that the body is hotter than is good for it, it sends signals to over two million holes scattered across the skin. The only parts of the body without any sweat glands are the tip of the penis, the clitoris, and the inner labial lips. American whites and negroes have the same number of sweat glands of course, while all right-handers are likely to sweat more on their forearms than left-handers, though each group curiously enough sweats more on the arm other than the dominant one. The brain's nerve signals cause tiny muscles to

Delight on a summer day. The principle of heat loss through fluid transport is the same here as in sweating, but the greater velocity and colder temperature of the hydrant's water increases the effect.

squeeze down on even tinier rolled-up tubes that conveniently form the bottom of these holes, and which, even more conveniently, have a few drops of a mostly water fluid waiting with them. Each tube is so tightly coiled that it would stretch four feet if unraveled. The squeezing presses the water out, and as the holes are very narrow ones, the water rushes up in a gush. That's two million miniature geysers, each one just a mite too small to be seen without magnification. As an added microcosmos glory, each sweat gland fires on its own separate schedule, having only enough fluid stored inside to erupt once every nine seconds. That produces an atonal symphony in geysers, an incredible landscape of uncoordinated spurts — first here, then there, then in between or somewhere else, repeating from these miniature duct-works all the way across your skin. It's no surprise then that the reservoirs of water left behind in your body get seriously depleted as you sweat.

Since the average person has about eighteen square feet of skin (about the size of a large breakfast table), that comes out to seventy-seven sweat pores on every square inch. But these are not evenly distributed, there being many more, for example, on the palm than on the elbow, a fact that gives justification to the telltale clammy palm so dreaded by those in the Mafia called to square accounts with their bosses. Half the total sweat comes out on your chest and back, a quarter on your feet and legs, while the rest is left for the head, arms and hands. For some reason more always comes out on the front of surfaces than the back, and sweating gentle or frenzied is likely to be first noticeable on your forehead, neck, chest or the top of your lip. In infants the glands are evenly spread across the body. But, like brain cells, once lost they can never regenerate. With time there is a selective dying-out that results in the adult distribution giving these effects.

Once any sweat fluid reaches the surface, it's ready to be evaporated away by any passing whiff of air however faint. The evaporating water takes heat from the supporting skin with it, so cooling down the skin as it goes. Sweat is almost entirely water, but as the sweat glands are not passive things, but actively screen through the blood to pull out just what they want, there are a host of special additives too. Your apparently simple sweat also contains salt, potassium metal (the seventh most common element in the body), milk sugars, and, as an added treat, the little-hailed substance bradykinin. Bradykinin clambers onto all the small blood arteries and tugs them open wide. It is one of the most powerful artery-openers known and is also a main part of the poison in a wasp sting. The wide-bore blood vessels it produces carry more hot blood up through the skin surface for cooling. As the skin contains a third of the fresh, hot blood from the heart, a little cooling there can do a lot. That's the crucial backup mechanism that keeps us cool: water and a natural wasp poison in it geysered unevenly all over you!

During heavy exercise, when someone might generate six times the heat he would if just resting, up to two gallons of sweat, a whopping sixteen pounds of water, can be spurted out into the atmosphere, cooling a bit each time it takes off. The energy increase during exercise is unique. Almost no extra hot blood pumping is needed when your body forms urine, secretes enzymes, or synthesizes sugars in the liver. But start a bit of jogging, or dash suddenly up the stairs and your heat output can suddenly be eight times higher than it was just before. Marathon runners have ended up with a temperature of nearly 106 degrees F. by the end of their slogs. It's no wonder that athletes get so thirsty. And as low humidity makes it easier for sweat to evaporate the same goes several times over for those in the unfortunate position of having to do forced army marches in the desert. At this extreme the body's sweating will be disposing of up to 90 percent of the internal heat, instead of the mere 15 percent it usually handles.

Sweat has no odor as it comes out of its tiny pores. But in the armpits and genital region, and a bit of the scalp, there are additional glands that produce dribbles of a milky white goo. (There's a third set of sweat glands inside your ears, but all they produce is wax.) These dribbles dry into shiny droplets that look like miniature balls of glue, and each one has some pungent ammonia traces locked inside. If the white glue-drops stay around, either through lack of airflow, or especially through clinging to the hairs in the armpits and genital regions, they begin to get munched on by the swarms of bacteria that are always waiting around on the skin. Once munched up, the globules really do produce a smell. That's something now less useful than it once was.

Drawing of man holding his skin, showing its large surface area.

A good strong stink could once have helped a straggler find his tribe in a thick forest, and in other circumstances a specific odor could have served as a reminder of the sexual delights to be savored in approaching closer to the source of the savor. All such acts are useful to the species, but now are so vestigial as to be just a nuisance. The sole exception are some of the sexual odors noticeable today, which serve however more as mild confirmation than wild initiation. Most pungent are the droplets from the armpits, least pungent is the outpouring on the scalp, while the odor from the pubic ones are in between (there are lots of reactive granules in the armpit droplets, none in those on the mons veneris).

In the front-line attack on these wet or bacteria-feeding activities of the body is the delicately named personal hygiene industry, which manages to sell over $540 million worth of deodorants and antiperspirants in America each year. It was not always thought a pressing need. Englishmen at the time of Shakespeare thought that excessive bathing impaired virility. One member of Parliament played it so safe that he was able to announce going 13 years without bathing, and there were many others not far behind.

Antiperspirants work by swelling up the tiny muscles and blood vessels around sweat pores and the odor glands, causing the ducts buried a tiny way under the skin's surface to fill up with fluid in lieu of sending it out. If this fluid is not sent out we stay dry, even though the brain might be frantically signaling to the pores to squirt away because the body is too hot for its own good. This can produce a swelling feeling and a slight fever in anyone who has whacked on lots of antiperspirant in anticipation of a strenuous walk in warm weather when they want to stay dry. The same feeling can also arise in someone who has put on lots of antiperspirant in anticipation of a stressful committee meeting or an outing with a potential sexual partner whose intentions have not yet been made clear, and whose interest might be deterred by a sudden dousing from all the two million sweat pores suddenly switched on full blast.

This is curious. Why should such scenes of social anxiety make the body act just as it does with standard overheating? To find the answer it's worth taking a closer look at the stressful scenes. What you really want to do in a meeting where your boss is browbeating you is to lunge across the table and throttle him around the neck, while a rather similar lunge, though with quite a different ending, is what a romantic evening out is continually suggesting to our hormones and nerve centers. This is where the stress begins, for our body doesn't know that such mad lunges are frowned upon socially, while the most fully developed parts of our brain know it only too well. This leads to the body's getting ready for strenuous action, be it forceful or delighted, but not being able to carry it out. The getting ready means, as we've seen, that blood sugars are assembled by the crucial muscles, coagulants are sent through the bloodstream, white corpuscle production steps up, the spleen soaks up blood for future use, and powerful hormones are distributed throughout the body. But when nothing happens after all this fuss, the blood sugars, coagulants and hormones get broken down wherever they happen to land. That breakdown produces heat. So did the original production of the corpuscles and the swelling of the spleen.

Normally this heat would accompany strenuous action, but now when there is no strenuous action the heat gets produced anyway. Once that happens, cooling mechanisms come into action, and that action, above all, means perspiration. As we saw in chapter two the branch of the autonomic nervous system that switches into high gear during danger or excitement, also has leads to the sweat glands, switching them on too. There's a curious special case. If it should happen that you start to produce this body heat in anticipation of major action but suddenly realize that the situation is too far gone for any action on your part, then the sweating will start but the body will not have a chance to heat up. That is the typical situation of real terror, and the resultant cold sweat has often been reported.

Sometimes of course the much-anticipated action, be it in the boardroom or bedroom, that is at the core of this battle with the antiperspirant will actually happen. If the antiperspirants then continued to do their job, you would suffer from such heat production without enough cooling that you would instantly get an incapacitating fever. Luckily, in extreme circumstances enough perspiration will get out to cool the body, for the best commercial antiperspirant holds back only around 40 percent of the cooling fluid available. (Those are the rollons; sprays can hold back only 20 percent, despite what their sponsors might proclaim.) Only if you were covered with something like a thin but continuous layer of gold paint that blocked every one of the body's two million sweat pores would you actually suffer a fatal overheating. That's why in making the James Bond film *Goldfinger* the actress was never actually painted with gold on more than one side at once as will be confirmed by those who remember that the camera never panned continuously across her body.

Well before antiperspirants get this far they run into another problem, which is that the muscles and other paraphernalia around the sweat pores do not take kindly to being squeezed so suddenly. A few become sore, a few more swell, and the result is a noticeable irritation. This effect means that antiperspirants will do only a partial job in controlling output, even when it's not extremely hot out. For that reason they are usually mixed with a deodorant that serves as a fallback device for making at least some of the rich bacterial brew produced whenever a few of the white odor beads dribble out, mix with the ordinary sweat in the armpits, and stick. Many of the most popular commercial deodorants actually contain antibiotics to get at the smell-producing munching bacteria.

Usually the salt content of sweat can be ignored in all this. It's there and we don't worry about it, except when it might dry to form rings on our shirts. But for someone lost at sea there's a horrendous limiting factor to our salt sweat levels that comes into play. Whatever happens, your body is incapable of raising the proportion of salt in the sweat (or urine for that matter) above 2 percent; this means that no one can drink seawater for long and survive; as its salt content is a higher 3 percent and the additional 1 percent would soon build up to fatal levels in the body. That's why a shipwrecked sailor can go mad of thirst in the middle of the ocean. Reflecting on the glories of an albatross flying overhead would not really make things better, for the albatross has evolved special salt-accumulating tubes and ducts leading down from its nose along its beak, so that any excess salt it gets is neatly disposed of. Seawater is all that infuriating bird drinks.

Increased Heat: Consequences and Catalysts

This discussion of how the body cools off by sending out little geysers of fluid has only hinted so far at the ominous consequence of not providing enough of this type of cooling. For someone standing upright, the problem comes quickly and silently. On a hot day sweating will be accompanied by a widening of the blood vessels near the surface, and that expansion usually helps in dispersing heat from the hot blood. Sometimes, though, it has more insidious effects. About half the blood vessels in the body are in the legs, and normally the excess flow there on a hot day will come back to the heart by way of the veins leading back from the legs. But when someone is standing up too long, the veins aren't strong enough to send this excess cooling blood all the way up back to the heart. If the blood doesn't make it to the heart, it certainly won't get pumped on to the brain. And without enough blood in the brain, consciousness slowly fades away, leading to a prompt faint by the overheated party.

This is an occupational hazard of honor guards, as they stand at attention on the hot asphalt on an airport runway for hours on end, waiting in tight formation for the plane of an arriving statesman. Such as the Nixons, Reagans and Thatchers of this world. Without enough wiggling of their toes in their boots to pump the calf muscles and so the veins within, they will faint. It's a hazard that usually takes care of itself, however, for a fainting soldier will have no choice but to fall. Once he is flat on the asphalt the blood that was in his leg veins to cool him off will have no problem getting back to the heart and so on to the brain, for, going horizontally now, it has no pull of gravity to overcome.

If the soldier were surrounded by duty-minded comrades who held him up after fainting, however, then the blood would never have a chance to come back up from his legs. He would sink into a deeper and deeper unconsciousness, and, if held up long enough, would die. That's precisely the principle behind the ancient practice of crucifixion. In many of the hot lands bordering the Mediterranean, this practice of tying someone to a pole so that he couldn't fall down after he fainted in the heat was widely used as a cheap means of execution.

This unpleasant practice ran in precise phases. First the veins in the legs of the unfortunate subject would expand as his body automatically tried to cool off in the day's heat. Then, as the blood flow from these distended vessels slowed down, he would faint. After that, as a third phase, he would rest unconscious as the blood in his body gradually pooled in the legs, with less and less making it up the long pull against gravity to supply the brain. His face would go white and cold. This condition is called orthostatic hypotension (from the Greek for "upright-standing low blood pressure") and has been widely studied by modern physiologists, especially by those NASA specialists who designed the space suits for the American lunar and Space Shuttle programs. People in this condition can be unconscious for hours on end, as the blood supply to their brain diminishes, without suffering any permanent damage. Because of the low pressure in the veins of the upper body, the subject's heartbeat and respiration are greatly slowed and shallow: in experiments with this on animals, an electroencephalograph (to measure brainwaves) is sometimes the only clear way to be sure that the subject is still alive.

The best treatment for someone removed from an upright position after falling into this deathlike unconsciousness from crucifixion is to lay him down in an air-conditioned room, or, in lieu of that, in something like a cool, damp cave. Recovery will then take place quite naturally in hours, or in extreme cases a few days, for the coolness will tighten the bloated leg veins, and the horizontal position will ease the blood flow back to the brain.

While this particular overheating problem is now mostly of historical interest, other problems remain. If the body's heat goes up despite the best that sweating can do, one is likely to feel giddy and tired. This effect is called heat syncope. It can happen working out in a gymnasium, walking through a city crowd on a hot, humid day, or just entering a car that has been sitting for a few hours in the summer sun. The giddiness comes about because blood is drained from the stomach muscles, leaving any food resting there to ferment uncomfortably without being ground up and led on to its proper exit, the intestine.

The blood coming from the stomach wall gets conveniently shunted into the paths that lead to the outer surface, there to be cooled by a leisurely flow not far from the outside air, which though warm, is rarely hotter than the internal organs.

That rediversion occurs whether one wants it to or not, but the other symptom of heat syncope, tiredness, is not so compelling and is one that some victims try to fight. They shouldn't. When it starts to get seriously overheated, your body makes you feel tired because that feeling is just the thing to get you to go inside and rest, thus cutting down on the muscular action that produces heat and allowing what heat there is to disperse gradually. Blood sugars, oxygen and other nutrients are scurrying to the most important internal organs such as the brain and the liver, so that if anything worse happens, these at least will be protected. And when the blood sugars are there, they cannot be in the muscles normally used for movement — thus the tiredness.

But when somebody resists the tiredness then the inbuilt recuperative systems don't get a chance to help. What happens next, if the person stays out in the heat and continues the shopping, hiking or just showing off that produced the heating is called heatstroke. In this malaise the usual ways of regulating heat go berserk, just at the time when they're most needed. The body actually starts breaking down more foods, thus producing more heat. The most efficient cooling mechanism, sweating, totally stops.

Heat is building up inside, and with no sweating, and no rerouting of internal blood to the surface, the skin acts like a close-fitting plastic sack. Inside that, the body temperature goes up and up. When it reaches 105 degrees Fahrenheit the convulsions start, but luckily delirium sets in shortly after, so that the final stages are unnoticed by the sufferer, though they are only too apparent to onlookers. The temperature continues up, steam can rise in small amounts from the eyeballs, and a coma begins that continues till the body temperature has reached 111 degrees Fahrenheit.

At this temperature the body is overloaded. All chemical reactions are taking place at about twice the rate as at the usual resting temperature. Millions of nerve cells are firing and firing for no reason, massive amounts of proteins are being broken down and circulated under high pressure, and many brain cells are beginning to unravel. If heatstroke reaches this stage, death is almost certain. The delicate internal structures, the proteins and tissues and food by-products, have quite literally melted out of shape. Ten thousand people are estimated to have died in the great heat wave of 1966 in the U.S.

If a way could be found to keep the body alive at higher temperatures, then a very curious effect would be felt. The brain cells of anyone still alive at 165 degrees Fahrenheit would be running at 12 times their normal rate, for they double their activity with each twenty degree rise over the usual. In a lifespan of 80 years, someone with this temperature would experience what at our usual rate would be 960 years of thinking, observing and remembering. But that would only be if the cells continued to work at this higher temperature, and that is unlikely due to the meltdown at 111 degrees.

This disconcerting destruction at 111 degrees raises the question of why heatstroke occurs at a temperature so close to the normal one. Normal is somewhere around 98.6 degrees, and the heatstroke cycle begins a bare 6.4 degrees higher, at 105 degrees. It might seem we're precariously close to overheating with the next sudden jog for a bus, and then continuing on the cycle of fatally mounting temperature. It's even more disturbing to realize that most chemical reactions the body needs to have take place, such as the release of energy proteins in the muscles, or the firing of nerve cells across the body, work best at the high temperatures of heatstroke. (This super-efficient working was in fact the penultimate symptom.)

A good example of their efficiency at distressingly high temperatures is the breakdown of ATP, the curly looking molecule which is the body's main source of quick energy, and is stored conveniently at hand in all the muscles. To get some use out of it a small corner section must be broken off. This separated corner bit ends up energizing the muscles when they need it. The problem is that at what we hope is the usual body temperature, it would take about seven years for the necessary corner bit of ATP to be snipped off. At higher temperatures such embarrassing pauses would not happen. If the energy-containing ATP were dropped in the right fluid stewing at 115 degrees Fahrenheit, the necessary corner section would drop off in a fraction of a second. Unfortunately in the next fraction of a second a bit more of the ATP would come sliding apart, and before very long the whole ATP molecule would have come apart into its constituent pieces. The same would happen to any other body chemicals that happen to be placed nearby.

How can the body provide the necessary heat to get the neighboring cells and molecules from melting away? The answer is that we don't produce the necessary heat. Our bodies have a better way, and it rests on the fortuitous use of tiny chains of molecules called catalysts. These are continually sloshing around inside of us, so many of them in fact that if each were worth one American penny, our deposit value at the bank would be over $800 trillion.

Catalysts are there to speed up chemical reactions without recourse to high temperatures. Need ATP-energy for a quick stretch with the arm muscles? The right catalyst slides up, does its stuff, and there the broken-off fragment of ATP is, ready to go to work. With catalysts the necessary piece of ATP can be cut off even at normal body temperature in $3/100$ of a second, and with the catalysts Sugar Ray Leonard has going when he's throwing a left hook, even faster than that. The catalyst is like a bartender who never takes a drink himself, but deserves the praise of his customers for getting the shakers ready and putting the drinks out on the bar, so they can then get what they want, coolly and without overheating.

The catalyst that works on ATP is an important one, but there are thousands and thousands of other types of catalyst, all with their different tasks, and all constantly chugging away at the ordinary comfortable temperature inside our body. You have catalysts that digest starch, move electrical fluids, and digest proteins: others that keep nerves working, tissues building and fat absorbing; and yet others that destroy germs, color your skin and even help make DNA work.

The range that catalysts do their work in is what we have come to call "normal" temperature, the old 98.5 or 98.6. It's enough under the heatstroke temperature to keep our internal molecules from turning into something resembling guacamole sauce. And even though it might not seem enough under that dangerous temperature to keep us from concern, that gap is rarely breached due to the very ingenuity of the catalysts themselves. For they do their crucial job by having a precise and delicate structure, which would be impaired by any rise in temperature. They would be the first thing to go when temperature rises, and when they begin to go out of action the body conveniently slows down, and so cools off. That's another reason why we feel tired when beginning to get too hot — the catalysts are cutting out, so only through a foolish force of will can we keep up the activity that is producing the heat.

This also explains why fevers often go up and down in exhausting cycles, for the rise in temperature of someone suffering from a fever will cool off enough catalysts for his interior activities to slow down, which is another way of saying that he will cool down. But with the viruses of a fever-producing disease in the system, this cooled body will be acted on to raise its temperature, starting the cycle again (there are other factors going on at the same time of course). The slowing of the catalysts and resultant rest is masked. When the first American astronauts were being selected, one test was to be put in the "bake-box," a dry room where the temperature was kept in the region of 150 degrees Fahrenheit. John Glenn got one of the best scores on his test, for his response to the heat was to lean his head on his shoulder and go to sleep — the best way of all to endure heat fatigue.

Decreased Heat: Shivering and Fat

To guard against the other extreme, the body has a number of systems always on call to keep from getting too cold. They work outstandingly well. Look what happens when an overheated Ferrari is parked on the street on a cool day. Within a few hours every part of the car, from the blistering engine block to the near-scorching exhaust pipe, will have cooled down all the way to match the temperature of the surrounding air. But now look at what will have happened to an enthusiastic child with a penchant for such cars, who has stood next to the Ferrari and admired it for the better part of the afternoon. His temperature will be the same as when he started, up in the high nineties Fahrenheit — way above the heat of the air.

The Ferrari cools so much faster than the child because once it is parked and the engine is switched off it has no more internal source of heat. It's just a question of time until the heat that remains drifts away into the air. For the child, rubbing his fingerprints all over the car, things are different. His internal source of heat never stops going. Either food is actually being digested in his stomach and intestines, or food that has been digested before and is waiting elsewhere in the body in the form of fat or proteins is being broken down. Some of this

Thin blue line on the profile edges of these figures shows surface layer where heat loss is omnipresent.

digestion goes into the youngster's contortions around the car, his pointing, rubbing and clamoring. But for the most part the energy that is freed when the food by-products are used goes solely into producing enough heat to more than make up for what he loses into the cooler surrounding air.

From each beer you drink, or each steak that you eat, a full 80 percent of the calories go into keeping your temperature up at the comforting 98.6 degrees. Only the rest of the calories, really just a gulp from the beer or one big slice from the steak, are used for all other purposes. Wasteful it might appear but you could not live without it. As the example of the parked car shows, all our heat would be sucked away even on a not particularly cold day, if this diversion of food for generating our own heat did not go on.

The greatest temperature drop anyone has survived was a fall to 64.4 degrees Fahrenheit, in the case of a black woman in the United States who was buried in a snowdrift for several hours. Unconsciousness is always reached well before that point, for chemical reactions slow down with cold, and by 70 degrees they're moving only one third as quickly as at normal body temperature. In that languorous state, consciousness has faded out like a TV set with its electricity reduced

by a third. Even when your temperature only falls below 90 degrees you stop feeling the cold, and just as sweating stopped with the onset of heatstroke, so useful shivering stops here too. For most people the heart slows, goes into spasm and then stops by the mid-60s. By way of contrast, an outside temperature of 64.4 degrees is considered balmy enough weather for meandering strolls.

The body has a number of fallback mechanisms in case ordinary food metabolism can't keep up and it starts to get too cold. They're quite neatly the reverse of the cooling mechanisms that come in during overheating. The first one on this side is to stop any cooling mechanisms that might still be operating. The little passageways in the skin that sweat comes out of (*poros* was the Greek name for passage) close tight, so no more perspiration can trickle out for cooling that way. Blood that had been shunted from the hot inner organs to small surface vessels is reshunted back to a circuit that stays deep in the body where little of the heat will be lost to the outside. Even the small blood vessels that were near the surface (and there are about twenty-five feet of them under just one square inch of skin) shrink down lower into the underlying tissue, where the heat in the little blood they carry is less easily swept away.

If the weather turns even colder, these small vessels will become narrower still so less heat-carrying blood gets through. Good as a heat shield but it can be startling for your friends. Seen through the translucent skin, blood flowing near the surface is what gives Caucasians their distinctive pink, gray or olive color. When these capillaries close down to conserve heat, the skin's natural translucence comes out clear by itself, which produces the arresting sight of a white person looking, for once, truly white.

This first line of defenses will start going without our noticing it, and do so every time we walk into an air-conditioned building, stay swimming too long, or just take a stroll on a slightly cool day. If the body is exposed to even more of a drain on its heat output, then shivering sets in. In mild forms this is just the rapid involuntary tensing and relaxing of a few small muscles throughout the body. Each time a muscle fiber squeezes like this, millions of ATP molecules — the body's main source of quick energy — are broken apart, producing some heat, and the muscle fibers slide and rub tightly along each other, which produces even more heat. As additional heat needs to be generated, other normally coordinated sections of muscle begin to clench and unclench out of phase. Heat production is likely to go up to 500 percent for a few key minutes this way. In extreme cases the strong muscles in front of the ear will join in the act, and as these are the muscles that usually hold the jaw steady, their shaking will start a nice vibrato going in the jaw. This is the origin of the Middle English word for shivering, *chivere* which stemmed, quite simply, from the Old English for "jaw."

Newborn babies don't have the coordination needed to shiver, and the first three hours after birth a baby's temperature can plummet 40 degrees Fahrenheit from the cosy womb warmth they had been accustomed to. But if they can't shiver they certainly can cry, and a good bout of crying will speed up the infant's metabolism by almost 200 percent. The same rise occurs when adults cry too, by the way, which is why tears of rage or joy lead equally often to overheating and a dampened brow.

If the body didn't lose so much heat in the first place, there would be no need for capillaries and chattering jaws. People who are going to live in cold climates, you might think, would do better to be built with most of their hot blood flow packed way on the inside and only a relatively small amount of potentially heat-losing skin to be exposed to the outside air. You would be right. If someone who weighs 170 pounds is a bare five feet tall, then most of his muscle, tissue, and other inner devices will be lying packed tightly against other bits of muscle tissue and the rest. Only the relatively small amount that is in contact with the skin will lose heat directly into the air.

Now imagine that same 170-pound person stretched up to a gangly six foot five. A lot of what was once warmly tucked away will now be lying exposed to the suddenly increased expanse of skin, liable to have its heat drawn away by any cold winds. Any Eskimo so gangly in construction would be hard put to find enough food to keep his own body temperature up at the needed 98 degrees because of the continuous loss to the air that his great surface area would expose him to. Such gangly Eskimos do not exist, for they all have the genetic programming to develop as efficient heat conservers, stocky and strong. The opposite is the case in the tropics, where a limber, gangly build is just what's needed to get as many of the internal organs as possible near to the surface, ready to get some heat relief from any passing breeze. Thus, the slender Melanesians and Watusi. When tropic dwellers are short — think of the Pygmies — they're sinewy and thin to keep a similarly large proportion of insides near the surface.

As confirmation of the advantages of a rounded shape in the Arctic, cold-weather animals do tend to be big. Such heat-preserving specimens include the polar bear, the musk-ox, and the walrus. It takes a long time for the heat produced in the center of a walrus, where the crucial food breakdown takes place, to work its way out through a neat 800 pounds of blubber in each direction. Fat is a tremendous insulator — like double glazing but better. (Even the fat that gives female breasts their shape can be considered to be there largely to protect the delicate milk-producing glands inside from damage both corporeal and thermal.)

Northern animals also tend to be hairy (who could imagine a great woolly mammoth in the Sahara?), for the layer of air caught in a thick pelt will stay there in all but the strongest winds, acting all the while as yet another insulating layer to keep the precious heat from being sucked out.

This has led to some curious misconceptions about the relation of body hair and virility. A marauding Norseman with his thick red beard is thought to be more virile (literally, more of a man, *vir* being the Latin for the male gender) than a refined Chinese mandarin with his few delicate wisps of chin hair. But if gaining hair is a good way to keep warm in cold regions, losing hair is the analogous way of allowing better air flow and so more cooling in the hotter reaches. This achieves its peak with the elephant and rhinoceros, two almost totally hairless creatures whose masculine endowments few would care to contest.

For what it's worth baldness itself can be taken as a source of male pride, for it only happens when there is a sufficiently high level of male sex hormone in the blood, and apparently never arises in those unfortunates castrated before puberty. But lest this tempt anyone to ease the natural process along with a close cranial shave, remember that the brain contains a number of proteins that are sensitive to even the slightest changes in heat, and even a few tufts of hair on top will help to moderate any sudden temperature changes within. A fully haired person will have 120,000 hairs on their head — more in blonds, less in redheads. On any given day 90 percent of the hairs are in a growing stage, and will stay, slowly pushing out, for almost three years. Ten percent are resting, and only for those in the 100 days or so that they do take that breather, is the chance of falling out very grave.

This genetic difference between peoples habituated to cold instead of hot climates is mirrored in the tribulations of the two warring factions that are likely always to be with us, to wit, the fat and the thin. Fat people not able to emigrate to the land of the midnight sun will suffer from the fact that they keep in a vast amount of their own metabolic heat, just like the Lapp and the Eskimo, but live in a climate where this merely produces an extraordinary feeling of overheating on summer days or in hot rooms: situations where the skinny people are maddeningly at one with their surroundings. Very fat people have it even worse, for if one roll of fat rubs against another, any perspiration produced in the fold will not get to gracefully evaporate away, producing a nice cooling feeling as it ascends, but will instead lie there getting clammier and clammier.

If warm days just make things worse for fat folk, perhaps cold days could make things better. Keeping body temperature up on cold days means burning calories to keep warm, and if that fact could be used to lose weight one would have the ultimate diet, automatic, certain, and above all painless, for it would demand no unpleasant reduction in the daily intake. Such a diet could well be labeled the Thermic Cure. Books could be written expounding on its virtues. Revivalist meetings could be held. All would proclaim the cardinal idea, that those wishing to lose weight Thermically must merely abstain from wearing a sweater around the house in the cooler months of the year. That simple move would make the body turn about 15 percent more of the food it's given into heat than it usually does. And that 15 percent saving on each bite would translate, in an average day's eating of 2,500 calories, into a 20-pound loss in the spread from November to March. There would, of course, be fanatics, who insisted that to get the full benefits socks must be peeled off, too. And there would be conservative types, smugly content with rolling up the sleeves of their woolen sweaters.

But alas, all the suddenly cheerful chubbies would be harshly deceived. The body can regulate appetite so well as to raise intake by just

Fat cells. These bulging depots are each composed almost entirely of a single large drop of accumulated liquid fat. They are excellent heat insulators. The nucleus, prominent in most cells, is here squashed into only $1/40$ the fat cell's volume. The deflated-looking balls just left of the center are fat cells which have lost some of their contents, a process which can be produced by hormonal action signifying calorie needs elsewhere in the body, and which, continued long enough, produces slimming. The partially enmeshing fibers are collagen, which hold the fat cells in grapelike bunches.

the few extra bites daily that will offset this extra burning for heat production. In an average adult lifespan the body keeps to within five or ten pounds, on the average, out of a total fifty tons eaten. Fifty tons — that's half the weight of the Space Shuttle. This means that the body is keeping its weight to a .01 consistency, and with that finely tuned mechanism keeping fat people fat, there seems little hope of escaping. As a crowning irony, on cool days skinny people will be even cooler than the fat ones, so they will use up even more calories in barely noticeable but constant shivering and capillary constriction, calories that in the fat people are saved to produce, you guessed it, more fat.

Heavy people insist that they're born that way and do not become fat through excess eating. The untold millions of thin people long dead who chose mockingly to call fat people by the Latin word *obederi* which roughly means "to overeat" (and has since been anglicized to our scarcely less insulting "obese") thought otherwise. They must have believed that eating less would end the fat. Modern confirmation is that carbohydrates — sugar, potatoes, flour, rice — are continually used up in the body, with a large adult having less than a single pound of it at any time. Half a day after last eating carbohydrates these internal supplies run out and the body automatically begins consuming its own fat reserves to start an alternative glucose-conversion pathway.

That's only part of it, though. A closer look at the body's heat economy comes clearly to the aid of the bulky ones. The habitually stocky build of Eskimos shows that some people can be born with lots of fat cells, which are placed so as to readily swell and keep the hot flowing arteries well away from the surface of the body. Mixed in with the muscle tissues or elsewhere, they will never entirely go away even on the strictest diet. With the right microscope you can actually see distinctively shiny, bulbous fat cells, mixed in with the very muscle tissue of those people who are disposed to be fat. It's even been suggested that human obesity is a form of vestigial hibernation.

But even this proof of genetic predestination won't still the cries of "more willpower" that rain out from the smug thin ones when a suffering dieter is in their midst. At least fat people can bask in the knowledge that it was their own ancestors who went out with rosy cheeks during the last Ice Age in hunt of the fearsome woolly mammoth, while the predecessors of the fashion models and basketball players of today were probably shivering around a fire in the home cave, waiting miserably for a scrap from the hunt. Or if they weren't, they should have been.

Extreme Cold Responses

When it gets cold enough even those of a spherical silhouette will fail to keep the continuous heat loss at bay. The wrong clothing, or just spending too long outdoors will do it. The damage sets in for anyone who gets way too cold, and it always follows the line of least resistance. The final thing a cooling body does is constrict the hot, fresh blood vessels in its peripheral regions. This preserves the heat for the most important inner organs, but it does leave the no longer protected outer regions open to assault. Hands, toes, and earlobes are the places where the capillaries are most likely first to draw down from the skin, so it's those areas which get hit by dangerous cooling first.

During this first stage of increasing cold, one can clearly feel it becomes harder and harder to move the hands or toes — almost as if the motor ability were being caught and taken away by the cold. This is reflected in the very word *numb,* which comes from the Middle English for "taken" — *nomen.* What's really going on in that phase is that nerves in the area that normally carry messages to and fro gradually stop working as the fluids that they're filled with congeal into something like a chilly paste.

Then the real damage of frostbite can start. Without any feeling in the fingers, it's easy to leave them with inadequate protection in the cold even longer. Filaments in the muscles just under the skin freeze and crack, which isn't too serious, but the walls of the few blood vessels left in the area also freeze and crack open, which is. If heat is restored later, the blood begins to flow back into the now leaky capillaries, which means that it never gets to the areas of flesh where the damage was the worst. Without any blood going to them to carry oxygen, the cells in the areas beyond the damaged blood vessels will suffocate from lack of oxygen, just as an entire creature can. As they die they turn black and rigid. One might expect that it is not a pleasant sensation to feel big swaths of your fingers or toes die, as frostbite victims wholeheartedly concur. On top of that pain, the plasma and blood cells leaking out of the damaged capillaries will swell there into a discolored, sensitive blister. With this double surge of pain, it's no surprise that many people faint as the feeling comes back after frostbite.

The best treatment for frostbite is to do the warming up as gradually as possible, so that the capillaries that are only slightly damaged will not burst when the blood returns. Vigorously rubbing the sufferer's affected hands in fresh snow is wholly mistaken. This treatment painfully shatters capillaries that had only been weakened and can produce an excellent breeding ground for wet gangrene and clostridial infection; it can actually lead to the affected parts dropping off later. Unfortunately there are always people around who plump for giving the painful vigorous rub, on the grounds that any remedy that doesn't hurt cannot be the right one. Such good samaritans abound. Beware.

Although frostbite is unpleasant for the hand or foot, it can be a wise strategy for the body as a whole — an instance of the general stress responses in action. If the body does not have enough heat to defend all of itself from the cold, it's certainly better to offer the limbs for a sacrifice than the kidneys or the heart. Your big toe can easily sink down in temperature by 40 degrees Fahrenheit or more when it's really cold out and you've been walking in slush or, even worse, standing still. It would be fatal to your heart for blood of that temperature to make it back untouched all the way inside. To avoid this the incoming cold blood is channeled alongside the outgoing hot. The two flows exchange temperature: great for the heart that now gets warm, not so great but a calculated trade off for the nearly frostbitten toe that now gets more cold. This strategy of selective fallback, as in a besieged fortress, continues as the body gets colder and colder. Legs are likely to get frostbite only after the feet have, and arms only after hands. All systems begin to shut down in reverse order to their real importance. Insight into this action has led to a remarkable procedure in certain delivery rooms.

If a newly born infant looks as if it might be suffering from oxygen deprivation, be it a tangled umbilical cord or the inability to begin breathing, doctors will dunk the child in a tub of ice-cold water. Thankfully the already flustered enough mother is usually in no condition to see this. The reasoning is sound, however, for when hitting the freezing water, even the newborn's body will concentrate its warmest blood in the region of the brain. And that warm blood is also the oxygenated blood. This extra burst of oxygen to the brain might give the doctor enough time to figure out a way to get the infant breathing before the brain is permanently damaged by oxygen loss. People rescued from the bottoms of ice-cold ponds have also been revived without brain damage, sometimes even after ten or fifteen minutes' submergence, something which would have been impossible without the protective fallback of vital heat and oxygen resources to the brain.

Heat, Hairlessness, and Signs

That mechanism is useful and discreet. The same cannot be said for a different, and totally worthless mechanism for guarding against chills which we are burdened with. When a hairy animal needs extra protection from the cold it can tighten the small muscles at the bottom of each hair and so jerk that hair straight up. As this happens to whole regions at once, the result is a dense forest of hairs, where a lot of air

Successive closeups of a hair on a human arm. The raised muscle at the hair's base is clear in the greatest enlargement, and is what swells up to produce goose bumps. In creatures on whom bodily hairs are closer, that would produce a more effective insulation layer as the hairs lifted up like a dense forest. It also can be a deterring sight to predators.

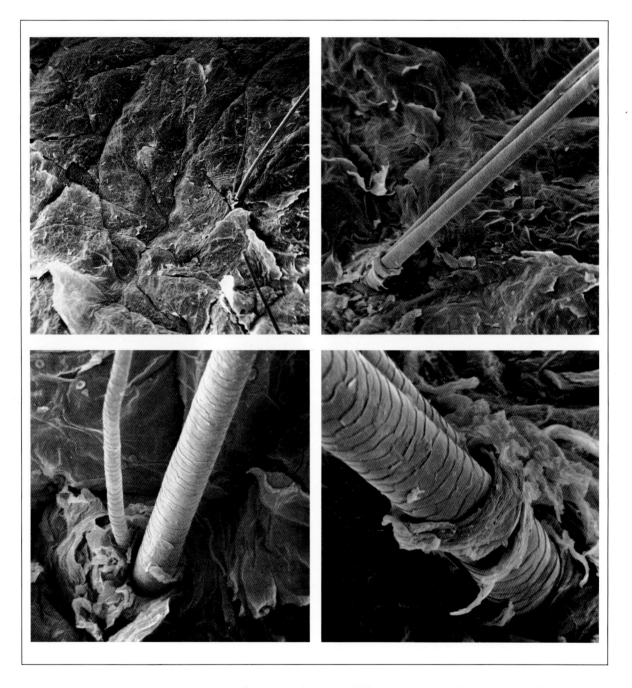

can be trapped, once cold but now warmed by the body's heat and held there as a captive to form a thick layer of insulation.

This sudden raising of the hairs can give the bear a noble appearance, or the lion a fierce countenance. Unfortunately in us humans, whose hair over most of the body has thinned to near-invisibility, the same tightening of the muscles around the hair roots in response to cold just produces a vista of tiny puckered mounds, as the flesh rises to support an erect hair that is no longer there to be seen. The whole

sorry affair is called producing gooseflesh, although why that worthwhile barnyard creature should be insulted in this way has never been
stated.

Our paucity of body hair does have benefits, though. Creatures all
covered with hair, from terrier to grizzly, may be spared the shame of
gooseflesh, but they're unable to give the delicate social messages that
changes in easily seen skin tone allow humans to give. These changes,
as we saw, are due to the body's main system for keeping its temperature right. When the small blood vessels in the skin relax and broaden,
then more hot blood flows through them to be cooled by the nearby
outside air. Through the outer skin's translucent covering, this provides a coloring change to pink in the case of white people, light orange for those with very yellow hues, and even a distinctive glow in
the case of blacks. Tightening these same blood vessels does the reverse. Then the body's hot blood ventures less close to the outside air,
and stays swirling around deeper inside, so conserving heat on a cold
day. A special case is that of white people who never worked, and
spent their days in chilly great palaces. Without much muscle on their
hands, and with only the cool oxygen-lacking blood of their veins
showing through their skin it would actually appear pale blue. The
dukes and lords of Renaissance Castille were such people, and it is
from their specterlike appearance that we get the term "blue-blooded"
for someone of noble demeanor.

The body will redden at any time when blood flows to the muscles
in anticipation of imminent action. As such it's a sure-fire sign of attentive interest, and as humankind is eternally optimistic will often be further taken as signalling interest in those nearby. This signal, intentional
or not, is called blushing. It's an especially common signal, too, for,
once started, it is hard to stop. It produces heat in the skin, so the em-
barrassment that produced the blushing will only increase when the
blusher realizes that the reddening signal is being given off. You can
almost say that we have eyes for heat in our skin: specialized cells in
the retina fire if triggered by electromagnetic waves shaking at 500 trillion times per second, and the brain registers that as light; specialized
cells in the skin of our face and arms fire if triggered by electromagnetic waves shaking at 300 trillion times per second, and the brain registers that as heat. All that produces more blushes.

Blushes almost always starts as two fine points in the cheeks, which
because of the distressing feedback from heat sensors in the area will
invariably spread to the rest of the cheeks. If it goes farther, as it so
easily does, the next place to glow tellingly will be the neck or the
nose, followed perhaps by the shoulders, the earlobes, and sometimes
even the upper chest. When it gets that far, the signal has heated up so
many muscles that they really are attuned for action.

It's possible that in a society of Amazons, daintified man would find
blushing the safest way to call for sex, so anything more direct would
be upsetting of the social order. But in our society, with phallocrats
still on top, it is women who blush more than men, and younger
women much more than older ones. The hot flushes that some
women experience during menopause do not fit in the blushing cate-

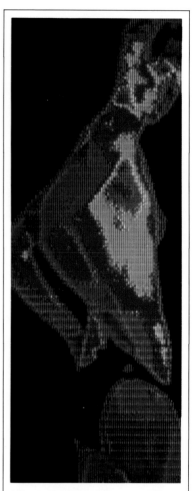

*The underpants the woman is
putting on are in fact white cotton. The thermogram shows
them as black because they
block so much heat loss compared with the uncovered areas.*

gory at all, though. They are not caused by shame at sudden infertility, as various male physicians have suggested in the past, but are perfectly respectable responses to temporary surges of estrogen, and other hormones, as the body's production adjusts unevenly to a new level.

The way the skin cools and flushes says a lot about the way clothes are chosen and worn. Clothes that cover large areas of body to keep it warm will not fulfill their function if they do not encode the same social signals that the body is used to making naturally. Some careful looking can almost always show how. When an Edwardian woman wore a tight blouse that would obviously be ruined by perspiration, she demonstrated that her husband was wealthy enough to keep her from having to work, and so to perspire.

There's a continuously adjusting balance between signaling social tags and calls with the skin, signaling them with clothes. It's an integral if often unnoticed part of all our dealings with the social world and can be ended only by a conscious choice to revoke those dealings, to turn within, and to enter the most personal of perceptual realms: a consciousness paying attention to its own consciousness alone.

This descent into the mind, accordingly, is the subject for our next, and final, chapter.

Regimented tranquility.

8
Relaxation and Sleep: The Day Ends

*T*his book began with the beginning of a day. Of course, no one is likely to have all the experiences described in this book so far — the rages and shivers, worry, birth, or burns — in the course of a single day.

But just as the very beginning of the day is the same for everyone — lifting the head up out of bed — so the very end of the day — going to sleep — is usually the same, too. And it turns out that the extraordinary changes inside the body when you sleep are related to the ones in a lot of other relaxing moments, from puffing marijuana, to sipping a Scotch and without the attendant problems.

To see how relaxations work, we need to begin with an important but rarely noted structure in the brain, and for that we need to turn to an odd corner of the globe and swoop down on a most unsuspecting creature that lives there.

Relaxation Controller: The Reticular Formation

On the eastern half of the Pacific island of New Guinea, nestled in the rain forests among the giant centipedes *Scutigera,* which crush large crickets with their feet, and the group-conscious spiders *Cyrtophora,* which attack in swarms of hundreds when a prey is caught and can pull down small trees with their webs, there lives a gentle-looking, almost blind creature with a long hairless snout and strangely splayed forelegs. This is the echidna, the spiny anteater.

Plodding about on the sodden forest floor, its tufted spines effortlessly keeping its singularly bizarre neighbors at bay, the echidna lives contentedly in a cozy world of its own, subsisting on a randomly browsed-upon diet of termites, insect eggs, and ants, all of which it methodically pulls in with its long sticky tongue. In its unhurried shamblings among the giant centipedes and team-killing spiders, the echidna enjoys the little-heralded distinction of being perhaps the only mammal that does not dream.

Not when the sun comes up, for then it's ambling about the undergrowth, nor when the sun goes down, as even then it keeps on poking about for grubs. Almost all other mammals dream, the lion and the cat, the elephant and man, but not the uncomplaining echidna, the most advanced creature in the evolutionary chain who does not. It will never, as the poet put it, experience "sleepe, the certaine knot of peace . . . the balme of woe." We, of course, are different, and the reason we can enjoy sleep and all the other types of blissful relaxation is that we, along with all other mammals have something that the forgotten, trundling echidna does not: an advanced, controllable reticular formation.

A reticular formation? This is a network of slender gray nerve cells, which are circuitously stretched to form a column of narrow vertical sheets in the lower reaches of the brain, just about where you massage the very top of your neck muscles. On their lower side they pick up sense nerves coming in from the body and on their top side they send these impulses up to the higher reaches of the brain. All our sense inputs go through it. But those cells don't send them up unchanged. There are too many messages coming in for them all to be treated by the higher centers with the same attention. A crucial bit of information could too easily be lost in the welter. Suppose an executive is receiving the following mix of information: "I see my desk is in front of me, I hear light music playing in my ears, I feel my stomach digesting lunch, I see that my secretary has gone berserk and is about to smash a swivel chair onto my head, I hear the light din of traffic outside."

It would not do for the brain to have no way of giving one part of that plan more attention than the others.

What the reticular formation does is to add to certain incoming signals additional ones that startle and grab the attention of the higher brain centers that the nerve impulses are going to meet. It's an intake censor: only those signals it "likes" pass in. Like a tuner on a radio, it improves the signal-to-noise ratio of information coming into your brain. The reticular formation ensures that the incoming signals *it* thinks are interesting enough get the crucial additions that will make the rest of the brain, especially the awareness centers just behind the forehead, perk up and pay attention. It has plenty of incredibly long stretching cells, that poke and weave and curl into almost all parts of the brain, to reach there. The reticular formation ensures consciousness.

That, of course, happens only if it deems the incoming information interesting enough. But how to tell? The reticular formation, after all, is just a dumb net of cells. At first it seems a problem. It's the very top of the brain that has all the reasoning powers, while the reticular formation down at the base is the one that has to choose what gets triggered into consciousness without itself being conscious, for if it too were conscious the problem would be removed one step to what makes the reticular formation itself aware of certain things.

The way the body avoids such time-consuming regressions is to stick an attention-getting "make this conscious" tag on almost all the incoming signals that reflect change in the outside world. This is easy enough to do automatically, because nerve cells throughout the body, including the ones in the reticular formation, are built to give their strongest message only when the impulse they're struck by is a changing one. That's how the reticular formation keeps us awake most of the time, and ensures that attention is focused on the most important things.

The reticular formation probably evolved this way as an energy-saving device. It's reasonable for a caveman to flinch when a tiger leaps at him and changes the image in his visual field, or for a computer operator to lurch backward when he gets a spark from a faulty connection on his keyboard. By contrast it would be exhausting if the caveman flinched every time he saw an unmoving rock or tree, and if the computer operator yelped and sank into a karate stance whenever he felt his hand lying harmlessly on his desk.

So how then does the full sway of consciousness get overcome? If there were no way to stop the reticular formation from passing on every new incoming signal with full, attention-grabbing clarity into the higher centers of the brain, there would be no escape from a tedious, possibly horrific, full consciousness. But there are ways to get at the reticular formation, to dull and sway it, and in the rest of this chapter we'll examine the range of ways possible, starting from the outside and working in. (Informal ways, such as just putting the feet up or taking it easy, produce their benefit in simpler manners, of course.) Trances, which hinge on getting the outside setting just right, will come first.

Trances

One of the most powerful trance techniques ever known was worked out by Ignatius of Loyola, one of the founders and the first General of the Jesuits, in his much vaunted *Spiritual Exercises*. Loyola had been a military man, if rather an unlucky one (he participated in one siege and was hit by a cannonball, before finding religion). His *Exercises* reflected the attention to drill that perhaps only a failed trooper can have. For four so-called weeks they were designed to last, four periods in which the Jesuit faithful were kept in an isolated monastery to prepare their minds through Loyola's instructions for the battle.

The key was silence. In silence the black-clad devotees woke up, in silence they prayed and ate, in silence they meditated, and in silence they returned to bed. Again and again till the four periods were up. All prayers were on similar themes, silently chanted, and constantly repeated. What this did to the initiates' reticular formation is easily imagined.

Consider how in ordinary life we sometimes drift off into a reverie when a colleague, boss, or spouse begins to babble on in a monotonous tone. What happens is that at first the speaking signal is enough of a change from what was coming in before that the inner part of the reticular formation sensitive to voices sends it on up to the higher part of the brain with the usual attention-provoking tag. But as the voice continues in exactly the same tone, the clarion call of the eternal bore, those particular circuits in the reticular formation begin to weaken and then to fade from overuse. That means that the consciousness-provoking signal is no longer added on as the voice impulses continue on their way up into the brain, and *that* means you begin to stare dully at the speaker with glazed eyes and fixed smile that suggest, quite rightly, that the body has been left on autopilot while the interested parts of the brain are off cogitating on other things.

In normal life only an exceptionally tiresome conversation partner is likely to strike your reticular formation into immobility like this so readily. Even then it will be brief. In the Jesuit activity, by contrast, the reticular formation is hammered out of action almost entirely. The lack of noise, the constant regimentation, and the drab clothes of all persons around the devotees cause the small number of nerve circuits in the reticular formation sensitive to those surroundings to fire and fire until they, and soon much of the rest of the formation, are almost entirely numbed.

It's close to going to sleep, but not quite. The reticular formation's residual wakefulness puts the body in a state in which it's especially sensitive to suggestion. This effect has been confirmed scientifically in modern laboratories, and more practically in a range of twentieth-century political interrogations. Notable examples were those questioned by the Russian secret police during the purges of the Old Bolsheviks, and American soldiers "brainwashed" by the Chinese during the Korean War.

In this respect Loyola was perceptively ahead of his time. A good way through the Jesuit exercises, just when the initiates were in their most dazed state, the silent prayers would switch from abstract ones worshipping God to more specific ones calling for action to defend the cross — action of the kind that Loyola wanted. The result was intended to create hordes of extraordinarily committed Jesuit "soldiers" to convert the world back to Roman Catholicism, and for a while it looked as if that might happen. For almost a century in the 1500s and 1600s these exercises terrified the population of Protestant Europe, for they were held to be the cause of the ability of the Jesuits, self-proclaimed soldiers for Christ, to torture, maim, cripple, and burn religious heretics, without suffering any efficiency sapping moral qualms. The efficient phalanxes were so committed to the instructions in Loyola slipped in during the mandatory exercises that the eighteenth-century pope Clement XIV ordered the dissolution of the Order of Jesuits when he thought the danger from Protestants was less than the danger from his own defenders. Other religious devotees have tried to produce conviction in numerous other ways. During the bubonic plagues of the Middle Ages the curious custom of self-flagellation, the beating of oneself with whips of braided leather or sharp razor bits, spread to tremendous proportions. Great crowds would assemble outside stricken villages, start self-flagellation and continue to do so in forced marches to neighboring villages until they dropped. They loved it. What was happening was that by irritating their open cuts the flagellants were likely increasing the circulation in their bodies of various well-studied chemicals, which are a product of suppurating wounds, and which in high quantity — as would be produced by the flagellators — build up in the nervous system to produce euphoria and trances. The flagellators' beatifically fixed expression, along with their pale complexion from the steady loss of blood, caused French peasants to call them the "gone over" people, gone over to another state. Instead of coining a word, the peasants used a variant on the old Latin root for "go across," *transire,* which is why this phrase, now at the base of "transition," is also at the base of our word "trance."

The second way of getting at the reticular formation by adjusting the setting is the easily induced, trancelike state known as hypnotism. It had a difficult beginning before the French Revolution in Paris, where an Austrian doctor named Franz Mesmer set up a "health clinic" on the Place Vendôme. The clientele were almost entirely women, wives of the wealthiest aristocrats in the kingdom, drawn by Mesmer's claims of using hypnotic forces to cure lassitude or ennui. The hypnotic technique involved seating patients in a cavernous, darkened hall, and having special assistants, invariably healthy young men, enter the room to effect the cure.

Silence was used, but for rather different ends than the Jesuits'. One assistant would sit before each noblewoman, place his knees around hers, and stare silently into her eyes while leaning close and massaging her shoulders, her back and her breasts. Mesmer himself would glide through the hall while this was going on. Kind patron, he was always ready to aid any of the aristocratic ladies who fainted when the "magnetism" got too intense and to help the well-developed assistants carry them to individual recovery rooms for more energetic curative work. Oddly enough, these occasional faints rarely kept the patients from returning.

Mesmer's curative center was closed by the authorities after a learned investigative committee was appointed to look into the goings-on and its members, including Benjamin Franklin, declared that although the practice did have a soporific effect on the ladies the possibilities for ill behavior were too shocking to allow Mesmer's clinic to continue. So it was closed, leaving only its founder's name in the term *mesmerism* as a reminder of what had been.

By the mid-1880s, however, hypnotism was back in fashion and by then perfected in its present style. The subject was told to concentrate on the hypnotist's soothing voice, or on a swinging watch — clearly a way to get the reticular formation zapped numb again. Hypnotized subjects take in suggestions with full credulity, not being aware enough to examine whether they really do want to stand up and sing like a bird, or say what they've always felt about their friends, as nightclub hypnotists are so prone to have them do.

In more serious efforts, hypnotized patients can appear to easily remember events from years before, but all too often the memory proves to be more imposing than accurate and is simply an attempt to please the hypnotist, who is greatly trusted during the trance. The trust, of course, has its limits. Total surrender has not been reproduced outside fiction. Yet the oft-mentioned claim that television is a form of hypnosis has some basis: brain cycles can slip into a regular 10 waves per second after half an hour of TV watching — a rhythm that EEG manuals say should occur only in someone not focusing his attention and "viewing a featureless illumination." Examine your viewing mates and behold.

Alcohol

Trances relax the reticular formation only if the outside setting is exactly right. Drinking alcohol is a way of doing it from the inside, and because of this it is more sure. The miracle drink that does this, or at least gives its devotees the feeling that it can, has been around for a long time. If you leave almost any fruit, vegetable, berry or honey outside in the summer, after a while yeast spores from the skin of the fruit will have reproduced enough to break down the sugar within and leave more powerful alcohol in its place. Alcohol is just sugar that has been eaten by yeast and then spewed back out again.

Midwestern farmers frequently benefit from the natural yeast effect and can sometimes be seen tapping the fermented juice that builds up at the bottom of their corn or wheat silos. Alcohol has been made at one time or another from such otherwise disparate delicacies as bamboo sap, millet, bananas, the fruit from the sausage tree, cactus, and wild flowers. The most popular sources of course are potatoes for vodka, grains for beer and whiskey, and grapes for wine and brandy.

Wine and the other alcohols owe their popularity to their being what the chemists call volatile. The word comes from the Latin for "fly away," and describes exactly what the alcohol molecule does if given half a chance. (Volley, as in the flying trajectory of a ball or the shooting discharge of an artillery volley, comes from the same Latin root.) Put a dab of butter on a saucer and it just sits there. But pour a half-glass of wine on a saucer, and its topmost molecules immediately start hurtling up into the air — an effect that could actually be recorded with a powerful enough camera.

In the mouth alcohol is no more restrained. Taste, swirl, and before you can swallow a sip, it has already started to evaporate into all the sensitive linings on the top of the mouth, and skedaddle around the back of the throat and up into the nose, where even more delicate taste receptors lay in wait. Even before you swallow a mouthful of wine, let alone a stronger drink, some of it has made its way through those thin linings into the bloodstream and is soon being perspired back out from the face and hands. You evaporate wine as you drink it, albeit only a little. Ten percent of ingested ethanol ultimately comes out in the breath and in urine.

Once swallowed, the alcohol does not get any less rambunctious. Its topmost layers of molecules continue to fly off, slamming into whatever is close. In the first holding station after the swallow that means the stomach wall. The zooming wine molecules stimulate it so that a wealth of closely knit blood vessels, which are usually closed

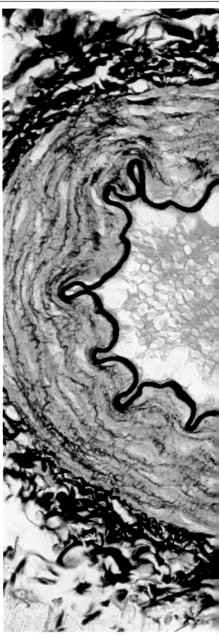

Inside view of main artery leading to the kidney. Alcohol not yet processed by liver will pass through here too without effect; breakdown products of alcohol, once metabolized, are screened after here and expelled with urine. Note musculature of channel, for varying pressure and hence speed of alcohol-infused bloodstream.

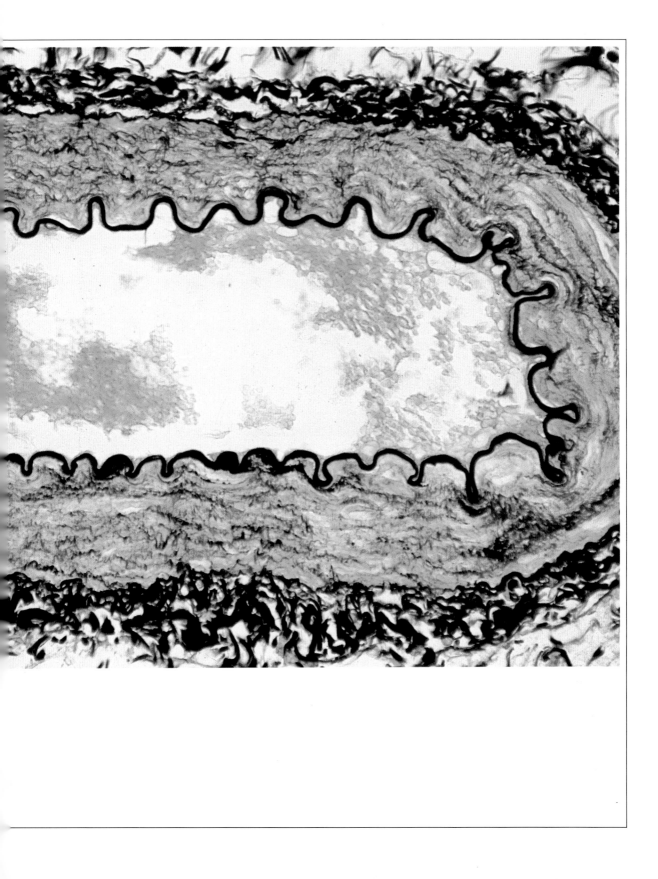

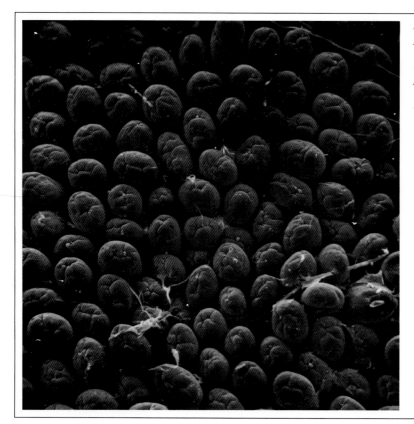

A field of nutrient-absorbing protrusions at start of small intestine, where most of alcohol absorption takes place. Each protrusion filled with a dense capillary system to transport absorbed substances into the rest of the body.

near the mucous surface, open up almost to the very top of the stomach lining, making it glow red. That way the stomach takes up the drink much faster than normal. Most alcoholic drinks send about 20 percent of their hard stuff into the stomach this way. Once leaving the inflamed chamber, the rest quickly slips on into the small intestine for the remaining percentage to be taken up by the body. When that happens, you get the real kick from your drink.

Left to itself, most alcohol will follow this route and be absorbed within half an hour. Numerous have been the efforts to slow it down. One of course is to put some food in its path. The best foods for this are those rich in fat and proteins, for they are the ones that best soak up lots of alcohol, and they also take a long time to get transferred on from the stomach. The Baltic habit of downing herring with your schnapps is not without its merit.

Some drinks have the reverse effect, and if anything speed their alcohol along into the small intestine. Champagne, really just an average proof wine, is known for its surprisingly quick punch, and that is due to the carbon dioxide bubbles in it. (All wine produces carbon dioxide when it's first fermented; in champagne the cork is just plonked in early on before the bubbles have had time to get out. The next Burgundy you drink once fizzed like the meanest soda.) Once in the stomach, the champagne's bubbles push hard against the narrow ring of muscle tissue that forms the opening into the next chamber, the small intestine. The compressed gas in those effervescent bubbles is enough to push that opening way open, much wider than usual. That gives the champagne's alcohol a free ride into the intestine, where the main 80 percent of the absorption will take place. That's one reason why champagne goes to your head so fast.

All the absorbed alcohol, whether from stomach or intestine, sooner or later slides along the hepatic portal vein gateway, which stretches under your ribs from the digestive tract into the liver. The first alcohol along this causeway gets the royal treatment upon arrival. It's skirted into tiny oblong chambers, and in the privacy within gets added pufferies through enzymes on store there that make it potent in a different way from the original: a "supermolecule." Yet as with so many stars, the moment of distinction is but brief. In .03 seconds it gets reshaped once more, this time in the meek form of a simpler and far smaller molecule. That takes less than another .03 seconds.

The end product of these two quick changings is far less agreeable than the original alcohol, being in fact the substance that makes cooking vinegar so sour — but it is quite useful though to the inside of the body. It gets no chance to wait around in its dressing chambers within the liver, but instead is switched into the usual metabolic channels for food. It might be turned into muscle, fat, or bone, but it is most definitely no longer alcohol.

That's what happened to the first drop of alcohol that was absorbed and got shunted into the liver. The same thing happens in quick .03-second flashes to the second drop, and even the third and fourth. But a bit after that the liver is full, and any more that's been imbibed has to wait its turn elsewhere. For liver has space to do the transformation of incoming alcohol at the rate of only about $\frac{1}{200}$ ounce each minute. That translates into a half-shot of whiskey or a small glass of beer an hour, and that's the only speed where almost no one will get drunk, or even tipsy.

Any drinking over that, sending in more than the $\frac{1}{200}$ ounce per minute, is going to change things. The liver can't handle it, and since it doesn't have any space for the excess to wait around until the dressing-change chambers are empty, it just sends it off into the body. Wham! That's pure alcohol being swept into the general blood supply, which means each heartbeat pumps it fresh into every organ. And in this alcoholic onslaught, the structure most readily affected is the brain.

The reason: the brain is an incredibly delicate organ. Although accounting for only 2 percent of the body's weight, it sucks in over 20 percent of all its oxygen and other energy supplies. These vast amounts must stream in constantly, for the brain doesn't have a single cell to store them in. A ten-second cutoff of the incoming blood sugar and oxygen flow will produce a total coma — and, after three minutes, damage that can be permanent. Because of this, everything that goes in has a precise role, and any newcomer would shake up the works.

That alcohol does; it's one of the few chemicals that can gush straight into the brain in its original, concentrated form, filtering through the normally selective barrier where the vast number of

blood vessels going into the brain (cut open a living brain and because of them it gushes) lead into the thinking brain cells themselves.
Proteins that are too big, minerals that are too "sharp" — all are
locked out at the final selection point where the membranes of the
brain cells act as a barricade. But not to alcohol. Its molecules are
streamlined to slip through this barrier. That's where the excess alcohol that the liver doesn't have time to transform goes — anything
above that initiates half-shot of whiskey or small beer an hour. And
once through the barrier, what are the first objects it meets? Thinking
cells in the gray matter of the cortex of course, but also none other
than the (our old standby) reticular formation.

When the alcohol hits, the reticular formation starts to fade just as if
the body has begun to go to sleep but hasn't quite gotten there, like a
television starting to lose its picture when the plug is pulled. It can
start with as little as .03 gms/ml of alcohol in the blood. At .05 the reticular formation is doing only half its job. It's still prodding the brain
into being aware of what's coming in, but it's only doing that with
something like half its usual oomph. A lot of sensations that would
normally be clearly noted — the raised eyebrow of a business colleague, the hesitant look of an undecided date — are now able to be
blissfully ignored. And with the reticular formation tagging less of
what comes in for attention, the field is free for greater attention to the
inner thoughts of the now suddenly carefree drinker, who is now most
likely astounding himself, and boring his fellows, with what he thinks
is the freshness and clarity of his thoughts.

This much will happen at just .05 gms/ml in the blood which is perhaps three long sips of a good scotch with ice, or two or three beers
in a quarter-hour. To the reticular formation, slowed by the overflow
from the liver, it's all the same.

Once started, more drink is likely to come. This was recognized by
the medieval Dutch, who had a harsh word for anyone engaged in
what they called guzzling too much liquor: *busen*. That term entered
English in the 14th century, passing through such spelling quirks as
"bouse" and "bowse," only to become the well-known "booze," applied to all alcoholic drinks itself, on the grounds that overdrinking
was sure to follow once some drinking had been started. And indeed
the cheerful consumer with his reticular formation dampened has always been tempted to keep the source of his delights flowing.

As even more alcohol comes in above the two-hundredth of an
ounce pittance the liver is putting out of the running each minute, the
level in the body's bloodstream goes up. That alcohol slips through
the fatty membrane of brain cells, sending the level there steadily up.
The effects being unleashed get steadily more potent to match. At .06,
hand-eye reflexes start slowing, and at .07, driving ability is measurably
down.

By the time a figure of .09 gm/ml is reached even stronger parts of
the brain will get enough alcohol to slow them. Most noticeable
among these is a thin arc of tissue across the very top of the cortex,
roughly in line with the swath covered when you push up your sun-
glasses. There's a central sulcus, or ditch, there, which is why the area

Brain's precise selection of incoming substance. View is from the top looking down, frontal lobe highest. In first picture, image marked "half-life" shows the speed with which different brain regions either absorb or reject druglike substance introduced in the bloodstream. Decision is quickest in frontal lobes (red), then rear of brain (green), then middle (black). Image marked "partition coef" shows regions where final proportion of the drug absorbed by brain tissue is the highest (red, yellow) vs. regions where the final proportion remaining in blood is greater (black, scattered green).

To indicate how precise such uptake is, the second picture gives the same images for a patient suffering senile dementia; differences are not great but serve as extreme outer limit of a normal subject's brain uptake range. That narrow variation in normal subjects is also the key reason why drugs such as marijuana and cocaine can be counted on to have the same effect on so many different people.

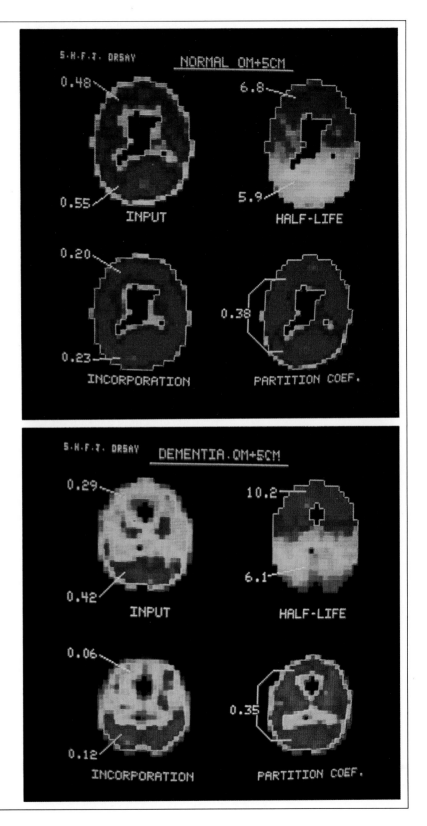

throbbing produced by a change in the balance of electrically charged particles in the blood and brain, and by the mere fact of having stayed up late a tardiness not generally noticed in the euphoria of the binge. Other specialists in the practice of hangovers will say that while the headache can always be cured with an icebag, the dry, sawdustlike mouth first thing in the morning is the worst part of it all. In cultural confirmation, they point to the French expression for hangover, *gueule de bois,* which means "wooden mouth." The thirst comes about partly because alcohol shakes up delicate filtration and absorption tubes in the kidneys, triggering it to send more fluid than usual out through the bladder.

More important, alcohol also messes up the body's habitual arrangement of fluids, which is normally neatly divided at 67 percent in the cells, 25 percent between the tissues, and the rest in the blood. As a total, water is about 60 percent of body weight; 100 pounds on a 165-pound man. The brain, for example, is 85 percent water, and even the heart is over 10 percent water. Alcohol shifts out into the blood or the space between the tissues a lot of water that's supposed to be in the cells. Sampling centers in the lower part of the brain, just inches behind the bleary eyes, pick up this lack and try to make up the difference. They send out chemicals that cause your body's cells to try to soak back the water they're missing. Unfortunately the mouth is not spared in this maneuver and as it holds the keenest taste receptors for waters you feel as if it's parched. If those sensors were on the back of your arm for some reason, that's where the debit would be felt (perhaps impelling the sufferer to roll out of bed, clutch his affected limb, and stagger down to the kitchen to get a nice wet drink to pour on it).

For some stout spirits however even morning-after parching is not the worst. "Drink a glass of water with lime," they might say (or "soak your arm in a glass of water with lime," if kind-hearted and not disturbed by the bizarre). What cannot be so easily remedied, they will insist, is the ultimate assault of the hangover, an assault that can bring down the mighty, crumple the courageous, and generally wipe out the legions. This is the queasy tummy.

The culprit here is again twofold. The flying-upward alcohol molecules that made the stomach glow bright red to take them up better is one cause, but the other villain is the little-known horde of additives that liquor manufacturers give their products, additions kept hidden by the special exemption they get from FDA's truth-in-labeling laws. Even a fine wine is likely to have, either as flavoring agents or as unintentional by-products of the fermentation, sugar, indigestible variants of sugar, dyes, molecular cleaving agents, ethers, alums, and aldehydes and lime sulfate.

The finest French wines also have — and this is perhaps the most rarely publicized fact of their whole legend-shrouded production — fishbones added in. Bitty ground-up ones but fishbones nevertheless. They're used to solve the problem of how to remove pieces of grape skin or other fragments from the wine, the stuff that otherwise would settle to form an unsightly sediment on the bottom. One way to do this is to pour in a bit of egg white, and let that make the sediment in front of it — the one concerned with motor control — is called the prefrontal region. This stretching strip of brain tissue is the place where many of the nerves controlling willed movement in the body start from. When the alcohol molecules start swirling in among them, getting in the way and slowing things down, its orders to the rest of the body are going to start with a bit of a garble. The rest of the transmitting nerve paths the orders take in the body may be unaffected, but that garble gets preserved all the way along to the end muscles: the drinker's arms and legs don't work quite as he wills, his hand gestures are overdramatic, his walking is a wobble. There's even evidence that alcohol enters right in and lowers the density of fluid in the semicircular canal balance centers in our inner ears — keeping the staggerer from realizing just how wobbly he has become.

At the same time, more alcohol molecules are ending up alongside the nerves that have received signals from the inner ear. Still as volatile as when they entered the body, the alcohol molecules bounce against these delicate nerves, altering the message inside them so much that less of an impulse gets through, and the whole auditory world seems to get gradually softer. Since at this time the whole auditory world is likely to be dominated by the proudly expostulating voice of the drinker himself, he will subconsciously step up his own decibel delivery, to bring the volume up to the level that he was used to before. Those nearby who are sober and hearing fine will find that their pleasant drunk has now become a loud one; those companions who have been drinking in step with the speaker are now likely to match him in loudness for the same reason. Drunks get loud, and many drunks get louder. People who have to get up early do not like to live near saloons.

With more drink, more effect. At .12 in the blood there's definite staggering, at .15 intoxication, at .25 nausea or stupor. Alcoholics are sixteen times more likely than average to die from a fall, and over four times more likely to die in a car crash. With .30 percent alcohol in the blood coma is likely to set in, and that stops the drinking right there. The amount of alcohol you would need in order to die varies greatly, but .60 gms/ml, the equivalent of 14 martinis in a quarter-hour is often enough to do.

Because unconsciousness sets in at half the fatal level, one might think that no one could ever drink himself to death. But that's ignoring the role of the stomach. Enough can be poured down the gullet *before* the blood level reaches .30 percent so that continued, and quite automatic, absorption after the drinker is out lifts it over the fatal .60 level. The Welsh poet Dylan Thomas is reported to have swigged numerous straight whiskeys with barely a pause during a lecture circuit in the U.S. he didn't care for, and the result was his demise that night. That is rare.

Hangover and Additives

After most drinking sessions there is a definite sign of life: the hangover. Connoisseurs differ on what is the most unpleasant part of this consequence for one's indulgent sins. Some opt for the headache, a

Erratum:
The order of pages 262 and 263 is reversed.

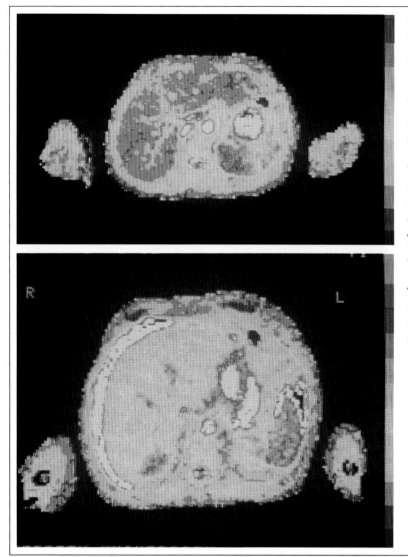

Cirrhosis of the liver. The liver is so crucial an organ that the body can usually regenerate it if even 75 percent were removed — but the fibrous and nodular distortion it undergoes if continued alcohol consumption is high enough will be too severe for even that regenerative power to help. Above, a normal liver seen in nuclear magnetic resonance scan. The liver is the blue structure largely on left-hand side. Orange, lower right, is the spleen, the circle in front of it is the stomach — glowing white from gastric fluid within. Separate circles on either side are the arms. Below, full-fledged cirrhosis, liver grossly swollen, heated green, the white arc to its left is the potentially fatal ascitic-fluid buildup inside the abdominal wall. The spleen is distorted too, affecting the body's germ defense.

pieces clump together so they can be easily strained off. In the higher class chateaus crushed-up bits of fishbone are used instead. They're mixed up into a gluey jelly and spooned into the maturing wine. Out comes the sediment, but not so all the gooey fishbones. Not a lot rests behind, but there is a definite amount, and a good protein analyzer will flash red in recognition if put near it. The fishbones don't leave much taste and only professional wine tasters can tell without machinery, but it can be a wonderful fact to drop when the wine steward is being more haughty than deserved

Harder drinks have more of these additions that never make it to the label — fusel oil and other delights can be naturally generated in the fermentations too! — with some being worse offenders than others. Vodka comes almost clean, having less than .003 percent of such additives on the average. That's why it's known by aficionados to produce one of the "cleanest," least debilitating hangovers. Bourbon, even of the same proof, usually has more than seventy times as many additives. Its whopping hangovers are notorious, and in those cases perhaps the only reliable solution is the favorite of all confirmed drinkers: more booze, to put the reckoning off to another day.

For some people drinking does more than just produce the occasional hangover, as something of a good-natured joke. For them it's more serious and alcoholism is the unpleasant but apt name. National variations in this are great. France unsurprisingly has one of the highest rates, at something like 9 percent of the population. The American number is about 2 percent, while Israelis have the lowest official alcohol consumption of any industrialized country, about .27 percent, drinking less than a seventh per capita the amount of spirits than do Americans, and barely a thirtieth the per capita wine consumption of the French and Italians. Heavy drinking can produce liver damage, brain damage, peripheral nerve damage, cardiovasculer damage, and a host of unpleasant social and personal decays. Why do people still get hooked? Sometimes their culture encourages it, sometimes they can't get enough of a break from ordinary consciousness through natural methods of sleep, well-rooted practices of religion, contemplation, or joy. There are, unfortunately, a host of other reasons too. Happy people rarely become alcoholics, curing unhappiness is hard, and with the enlightened exception of Alcoholics Anonymous, is often ignored in the treatment of alcoholism.

Harsher, strictly medical methods are far too frequent. Trying to cure the desire to drink specialists have gone at alcoholics with steroid injections, oxygen injections, antialcohol injections that make them retch, nicotinamide-adenine dinucleotide injections. They have subjected them to brain surgery, intravenous serum drips, intravenous alcohol drips; dosed them with strychnine, antihistamines, and tranquilizers. Few of these treatments have shown any effectiveness in controlled studies. For diehard drinkers, alcohol is the only way to get their reticular formation and thinking cortex in the condition they want it — down if not off — and until the world they have to face changes for the more comfortable, they'll keep on leaving it through drink whenever they can.

Drugs

Alcohol has its effect when an overflow that the liver can't handle wends its way to the brain. Mind-changing drugs work only in the brain, being made of cunningly shaped molecules that are used up only when they've been taken up by targeted receptors in the reticular formation and elsewhere. It's as if they get into the switchboard of the nervous system and sabotage it. And it's incredible but most of these mind-changing drugs are naturally grown by plants. The big ones — mescaline and opium, cocaine, nicotine, and caffeine — are well known. But there's also solanidine in potatoes, and tomatidine in tomatoes, both of which will have strong (and toxic) effects at suffi ciently high doses. These chemicals are plants' way of keeping away insects and large herbivores. One of the very few exceptions to this in-

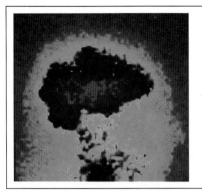

Right profile view of the flow of amphetamine from a swallowed capsule into the brain; marked with gamma-ray-emitting io-dine-123 to show uptake in higher thought centers of cere-brum (glowing red; green is the unaffected skull and face tissue).

tention is the alkaloid-loaded nectar that several plants pour into their own flowers — quite likely a reward for the bee, to keep it coming back!

LSD LSD is the most intense. It was discovered by a Swiss chemist named Albert Hoffmann in 1938, in his investigations of the various fungus growths that attack damp wheat and rye plants. The reason it developed in the fungus-infested rye, most likely, is that those growths developed LSD as a defense against people killing the rye by pulling it up and eating it. LSD is fatal in excess and there have been many cases where the rye plants and its fungus were able to keep down the num-bers this way of the humans so bent on attacking — i.e., eating — it. Some 40,000 people died in an outbreak of LSD-loaded rye fungus in the year 994 in the south of France; other successes on the rye's part have happened every century since (there was a big one in Russia 1926), with a notable example being the Salem "witches" of 1692, who seem to have just been peasant girls who tasted the infected rye pro-duced in the especially wet harvest that year, and naturally were thought to be under the Devil's hand when they started babbling on about the bright lights, weird visions, and other events of their undi-agnosed, and ever so unpropitious, Puritan trip.

Hoffmann noted this new chemical's structure, filed it, and forgot about it. Only six years later, in 1943 did he accidentally swallow a small dab. But for that, no one else would probably ever have heard of it. Writing later, Hoffmann reported that soon after his swallow he was overcome with strange, mysterious sensations. He didn't call them psy-chedelic, for that word was yet to be coined. What he did do was more telling: he went back to his laboratory flasks when his "sensations" ended — the first scientific LSD trip in history — and without telling anyone else, quickly fabricated himself some more.

At the end of the war Hoffmann published his work. It was quickly taken up by the U.S. army and the CIA as a potential, novel way of fighting. Their research got no farther than laboratory tests, which, leading to suicides by unwitting subjects, were not at first widely pub-licized.

Led by Aldous Huxley and others, the first popular spread of LSD came in the late 1950s and early 1960s, among, of all people, advanced-thinking bankers, East Coast lawyers, and even Henry Luce, mogul publisher of *Time* and *Life*. An effortless mind expansion, a pleasant novelty, and one not at all tainted by contact with undesirable characters. This exotic chemical was cheerfully sold by the Sandoz pharmaceutical company (its very interesting pharmacologically, and psychotherapeutic use was long claimed), which also marketed baby formulas and antibiotics. Only from the early 1960s on, when LSD's use had spread to campuses and reports of bad trips began to filter in, did it lose its appeal as a high-status drug for the elite, the super-cocaine of the day. Consumption seems to have peaked in the late 1960s. The head of the FDA estimated (perhaps excessively) that 15 percent of all American college students had tried LSD at least once. By the start of the 1970s it quickly began to fade into traditional hard-drug circles. The drop was helped by the report, probably false as it turned out, that LSD could damage a user's chromosomes.

That just helped a natural decline, for its effects were probably too strong for a student body that was graduating and beginning jobs. LSD was not a drug that could be taken, then ignored if the user wanted: it forced strange pictures on the mind, a process the Greeks called *phantazesthai* and we call "fantasizing." (The same Greek word came down into the twentieth century in a most distinct other guise: the title of the Walt Disney children's film, *Fantasia*.)

Above all LSD commands attention. This should make us suspect that the reticular formation might be involved here, too. Neurochemical studies suggest that indeed it is, for LSD is shaped almost exactly like serotonin, the chemical that we shall see is produced in the brain at bedtime, and that cuts down the powers of the reticular formation to make the rest of the brain pay attention to the sensory impulses that are coming in.

The details of its working though are more complex. One might think that because LSD is a chemical near-twin of serotonin, it can fit wherever serotonin can, and in particular could lock into the little enzymes that would usually be free to break down serotonin after it has done its stuff. With that regulator out of the way because of the LSD slotted into it, the natural serotonin that the brain is always producing would quickly build up. However that vague picture of LSD sloshing around is not good enough: even a hallucination is an incredibly elaborate, highly differentiated phenomenon.

One of these detailed actions is that the reticular formation gets stimulated in ways it was never designed to be. Nerves that are really receiving no signals from the body's sense receptors are blasted with excess serotonin molecules. That doesn't make them fire as if they are getting these sense signals anyway, but it does likely open gates too the memory centers. Firing of cells there gives the vivid impression of a lot of activity in the body and beyond, activity that is really nonexistent but that the brain has no way of disbelieving.

Above all, remember how intensively selective normal brain activity is: how the reticular formation must help in rejecting an immense amount of potential material all the time. Interfere with this sieving, even a little bit, and you naturally account for the instrusion of imagery, the mismatch or intensification of signals that people on LSD report. You also account for the whirring impairment of consecutive thought and memory they report — thoughts and memories which normally depend on the continued adherence to a given choice of what to pay attention to.

This effect must have been impressive to those who experienced it, but one can't help wondering, as Thurber did about the violin-playing seal, of what use it is. The question apparently was a common one, and the user of LSD may have died down, for all its splendor, not just because chromosome scares, or even the need for students to enter the labor market, but because of simple boredom.

Heroin

Of the other drugs, heroin must be the nastiest. It is not nearly as mind-changing as LSD — maybe just a bit of daydreaming is produced — but it does change *mood,* producing a euphoria so enticing that addicts will ruin their lives to get back to it. Heroin was actually first produced by choice, in 1898, in a committed attempt to make a nonaddictive version of opium. What a mistake! It turned out to be several times more addictive, for heroin, and, to a lesser extent, opium and morphine work by blocking the transmission of signals between certain brain cells. This makes the nerves much less sensitive to the usual chemicals that squirt between them and help carry the signals, and that, for anyone feeling dispirited or in pain, is a wonderful high by contrast with what came before. All of a sudden, just after injecting a single dose, all your worries and concerns, all your knowledge that there can even be worry and concern, is gone.

If heroin stopped there, it could be all right, but it doesn't. After a few hours the brain tissue adapts, and becomes less sensitive. As long as there is still some heroin in the body, it's a stalemate; the high might be over, but the taker doesn't feel especially bad, either. But once the heroin has worked its way out, through the ever-vigilant kidneys and liver, then the brain cells switch into hypersensitivity. Stronger signals than before flash across nerve junctions everywhere, including, to the junkie's distress, the very grouping of brain cells that was registering his pain or creating his low spirits before. If he felt bad before, he's going to feel worse now. A lot worse. To end the discomfort he'll need more heroin. What that means is seen only too well in the word that describes this state: *addict* comes from the Latin *addictus,* meaning "bound" — like a slave is bound to his master. That, to the eternal chagrin of the English chemist who made it as a cure in the first place, is just what happens with heroin.

Marijuana

Perhaps the most popular mild drug other than alcohol is marijuana. It can be a mild hallucinogen, but it is nowhere nearly so potent as LSD, and it can be a mild relaxant, but it is nowhere near so brain-charging as heroin. Its journey through the body is extraordinary.

Lungs developed from the buoyancy bladders used by certain early fish to keep afloat in the water. Lungs have no musculature of their

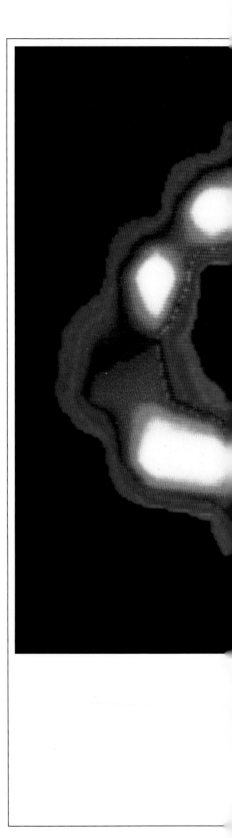

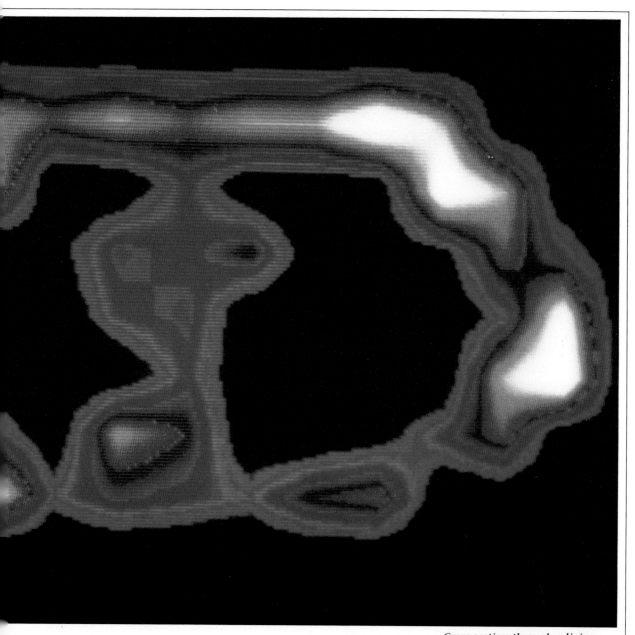

Cross section through a living chest, showing the great cavity that the two lungs fill, and the muscled wall around it whose sudden movements for respiration help suck in or force out air from lungs. The image has been produced by measuring the radiation produced by the body's hydrogen nuclei subjected to an intense and varying magnetic field.

This is the windpipe interior — purple-red fronds are cilia, yellow spheres are mucus globules being rolled along up and away from the lungs by cilia.

Blood flow dividing to reach all the minute cavities within the lung; at each extremity (and even smaller ones not visible here) potent chemicals in inhaled marijuana enter the pulmonary vein leading back to heart, thence into general circulation. Ribs are visible in foreground, especially at top; the segmented column at right is the spine; the diaphragm muscle, which distends to produce inhalation, is visible at the bottom.◄

own that can suck in air, and so by themselves they could do nothing to help you inhale marijuana's smoke. Instead, the large curving sheet of muscle just under the lungs, the diaphragm, pulls way down, while a number of slender flat muscles between the ribs around the lungs pull out. The result is a large space suddenly opening up around the lungs, which they're quick to fill, and it's that jerking pull sideways and down which we experience as a breath.

Down toward the lungs the marijuana smoke goes, almost lost in its descent among the other inhaled substances that come in, too: nitrogen (80 percent of each breath), oxygen (18 percent), carbon dioxide (.03 percent), and as an extra addition for every city-dweller, some 20 billion particles of grit, smoke, and mineral each day. In the marijuana puff there is also tar, carbon monoxide, fungi and all the other delights of any cigarette smoke.

Not all of this gets full down the windpipe into the lungs. Many of the grit particles land and stick onto the layer of mucus lining the inner surface of the windpipe. That layer is continually being swept upward at a rate of ⅔ inch per minute. Once it's up at the back of the mouth it will be disposed of either with a phelgm-full cough or by an innocuous swallow, either way getting it down safely away to the stomach. The marijuana molecules themselves are too small to account for the frantic coughing first-time smokers often go through as they try to inhale the drug in relaxed, elegant fashion. What's happening here is that marijuana must be deeply inhaled for the potent THC in it to get to work and that inhaling is more than most debutants can handle.

The marijuana that gets past is whisked through the main branchings into the lung, heading through smaller and smaller mazes until it arrives in the smallest of them all, the minute chambers called the alveoli. The lung is so riddled with them that their total surface area would be over 800 square feet if pulled out and spread flat — 30 times greater than the body's surface area, or enough to drape over a good-sized garage. There are over 300 million alveoli in all.

There can be so many packed in there because the alveoli are thin. Each one has a wall just 1/25,000 of an inch thick. That's all that separates the tiny air volume they encompass from the rushing blood cells pumped through from the heart. The marijuana slips right across, and gets stuck to the protein in the blood.

The amount of blood in all the alveoli in the lung would fill a cube five by five by five inches. It all gets permeated by the inslipping smoke. There's a fresh volume of blood filling up the alveoli walls every three-quarter second, and that loads up the whole bloodstream quickly. Smoking in the morning will get marijuana across the alveoli faster, for the bronchial breathing tubes peak in width at 10 A.M.; holding an inhaled puff will also increase the uptake, for it gives marijuana molecules time to slip into more of the quickly changing alveolar blood.

Once in the blood, the marijuana molecules go everywhere the blood leads — liver, legs, elbows and whatnot. Above all, the marijuana ends up in fat deposits all across your body. That's because cannabis produces an intensely fat-soluble molecule. The absorbed marijuana molecules are voraciously absorbed by all your fat stores, from the often noticeable ones in the rump and hips, to the less obvious ones around the kidneys, in the mesentery membrane supporting the small intestine. A single strong inhalation will send marijuana into the soft lipid material in them all.

After achieving this peculiar dispersal the main bulk of the marijuana can gradually leach off to affect the brain. (Some went there in the first swirl around the bloodstream of course.) Being fat-soluble the marijuana molecules are taken up by the brain cells' membranes, possibly by fitting selectively into certain spaces there that are just the right shape for them.

It takes a while for an appreciable amount to get to the brain from the body's fat deposits, which is why marijuana often takes a while to start acting. But it is also likely to continue heading up to the brain in a steady release once it has begun, which is why the effect can linger for several hours. (Eaten rather than inhaled, marijuana takes even longer to reach the fat despoits thus prolonging both stages.)

In the brain, marijuana acts analogously to a low dose of anesthesia. This can impair short-time memory: if something is said to you, usually you can concentrate on it by excluding other incoming material — a task involving the reticular formation. But with the brains usual sieving of excess material thrown out of kilter you get the typically good-natured yet vacant responses of marijuana smokers.

Should you take marijuana? That's a matter for personal choice. But there are several powerful reasons against doing it more than occasionally. Smoking marijuana sends as much cancerous tar into your body as ordinary cigarette smoking does. Anyone who avoids cigarette smoke at the office and on trains is being inconsistent, and unkind to their lungs, if they light up back at home in the evening. Marijuana also slows your reflexes — just about as much as heavy drinking does. That makes someone on a marijuana high likely to be just as deadly behind a car wheel as a drunk. (Marijuana stays out of the traffic death statistics, pharmacologists are certain, only because there's no such convenient blood test for marijuana as there is for alcohol.) Above all, a society in which marijuana dominated could not be expected to be an especially thoughtful or successful one. Arab countries which turned to it in the centuries since Mohammed outlawed alcohol are perhaps a case in point.

Not only is practical effectiveness reduced, but even the fresh insights that can seem so exciting when one's in a different mental world with marijuana or other mind-altering drugs, are likely to look dim in ordinary life. An illustrative case is the one in which Winston Churchill is said to have woke up suddenly in the middle of the night convinced that he had just undergone a profound, mystical experience. Luckily he had writing paper and a pen by the bed, so scrawled down his insights before falling back to sleep.

The next morning Churchill was eager to see what revelation he had had during the mind change, and read in his own handwriting on the bedside paper: "The whole is pervaded by a strong smell of turpentine." This does not strike the nonbeliever as an especially profound observation. The conclusion perhaps is that the world of hallucinations (Latin, *alucinari,* to wander mentally), as of all mind liftings we experience, should most definitely be kept as a world apart.

Sleep

Finally, the universal relaxation state, where the body manages to turn off the waking system located in the reticular formation by itself at the end of every day. It's a process so gentle that when it happens it seems we're effortlessly slipping into another state altogether. It's a state that got its name in England in the Dark Ages from a variant of the word "slip," *slepe,* or as we call it now, *sleep.*

We fall asleep when a group of cells smack in the middle of the reticular formation — right where its netlike vertical sheets blend — dribble from their centers a dose of serotonin molecules: the chemical that we've seen is curiously related to LSD. This serotonin soaks slowly outward to collect on cells throughout the reticular formation and, slowly, but unhesitatingly, damp their firing to a stop. Once that happens, you're out. Asleep.

But how to make it happen? We can't very well say "listen central reticular structure, release your dampening chemical." It's just not going to listen. The old adage of having a glass of warm milk before bedtime has a certain scientific support, for milk contains a good deal of a chemical which is a precursor of serotonin. But that chemical in

milk can take a long time to diffuse to the brain, and milk probably
works by the more comforting effect of a maternal warm drink. In a
suitable laboratory there would be no problem in causing sleep. The
reticular center could be triggered to submerge the rest of the formation in an artificial gush of the special chemical that turns it off. But
that takes the right equipment, microelectrodes and the like, which
are not normally found in a typical bedroom.

In lieu of that direct method, vast numbers of more subtle techniques have been developed to bring about sleep. They all hinge on
trying to reproduce the boring setting that would produce an unintentional sudden doze during the day. Although that is really just a feeble
attempt to damp down the reticular formation a bit directly, and hope
that makes it easier for the sleep-inducing knot of cells to send out
their rather more effective chemical load, it's the best most people are
able to do, and it's what we're brought up to think is the normal thing
to do before going to sleep.

In almost every culture people getting ready for sleep turn down
the lights, go to a quiet place, and align themselves in a position
where they won't have to move. That's fixed. While advocates of sleeping clothes or no clothes, water bed or spring ones will argue eternally over the merits of their favorite in hastening the process, once it
does happen the body is asleep, and a whole new world of happen-
ings begins.

At first it's barely a change at all. If you gently call the name of
someone who's just gone to sleep, his heartbeat will speed, and he'll
begin to sweat slightly — all without waking up. In that lightest phase
an adult's brain waves slow down to the typical speed of a seven-year-
old's waking brain waves. This is interesting because seven is just
about when most children first become able to keep clear logical notions, such as that the amount of water in a wide jar doesn't change
even if it's poured into a long, thin glass. Before seven they can't. This
revocation of the fully rational is just what might be expected for
someone heading on the strange descent to dreams, and it's pleasing
that the brain waves back it up.

The first stage of light sleep is also when there are likely to be
those convulsive leg-seizing twitches, so startling to bedmates yet unnoticed by the sleeper culprit. They occur when nerve fibers leading
to the leg, in a bundle nearly as thick as a pencil, suddenly fire in unison. Each tiny nerve in the bundle produces a harsh tightening of a
tiny portion of muscle fiber that is linked to it down in the leg, and
when they all fire together the leg twitches as a whole.

This lightest phase of sleep lasts only fifteen or twenty minutes.
When this phase comes again later in the night, there is unfortunately,
a good chance it won't come in peace. Linguistic philiacs might delight
in bruxism (tooth grinding), somnambulism (sleepwalking), and som-
niloquy (sleep talking), all of which are common during light sleep,
but it's doubtful if anyone, except the initiator, has ever delighted in
snoring. The horrenduous thunder-shaking of the softer tissues at the
back of the throat measured at up to 69 decibels — almost a pneu-
matic drill — has been indulged in at full tilt by Mussolini, Theodore

Roosevelt, King George IV, Churchill, and untold millions of blissful if little-loved others. Children, cats, dogs and hogs are noted snorers, with horses being capable of the feat while either standing or reclined. Sleepwalking can be more entertaining, at least to those of an adventurous bent, for not only will up to 5 percent of all children do it at one time or another, but jockeys have been reported riding during it, swimmers swimming, toll payers resolutely demanding their correct change, and quite likely on the appropriate day of Christmas, fiddlers fiddling.

All during this sometimes so peculiar lighter phase, the crumpled skein of cells that make up the reticular formation down under the brain is continuing to be squirted with the special, LSD-resembling chemical from the central cluster that started the whole process off. The reticular formation becomes more and more numbed, and soon even a loud calling of the descending sleeper's name won't produce even the quickening pulse or light sweat spurting it did before. More chemical, more numbness, and then usually about seventy minutes after going to sleep, the most solemn level of deep sleep is reached.

At this point your heart is slowed, all your blood vessels are sagging inward more than they ever do during the day, and the long stretching heat sensors in your hand are blunted and unresponsive, while the spherical multilayered cold receptors next to them are tightened down, like an onion trying to shrink into its core. In the top of your nose the uneven, almost pitted terrain where smells are identified is barely working, the bones that conduct sound in the middle ear are stretched as loosely, and ineffectively, as they can be, and the tiny protrusions on your tongue, shaped like miniature mushrooms, that normally detect taste, are still and so dry they wouldn't even squirm before a lemon.

All the senses are closed down to the deepest notch possible, the brain cells are firing in ultraslow, strangely synchronized patterns, and the body is ready for the next step away from ordinary life, the jump into what the medieval French called a "wandering," and what their older English counterparts had called "joy" and "music," or as we put it, "dreams."

Dreaming

What is in the dreamworld when you get there? People mostly, including in almost all instances, the dreamer himself. The consistency across large surveys is distressing to those who pride themselves on the originality of their most personal inner views. Here are the median figures to see where you fit in: familiar characters are dreamed of 45 percent of the time, unfamiliar 55 percent. Men are twice as likely to dream of men as of women, and the figure rises to almost 3:1 with advancing age. Crowds and groups are dreamed of 30 percent of the time, individual characters 70 percent.

Politicians are almost never in dreams, even during an election while the dreamer's own family steps into about 20 percent of dreams, with mothers showing up 34 percent of that time, fathers 27 percent,

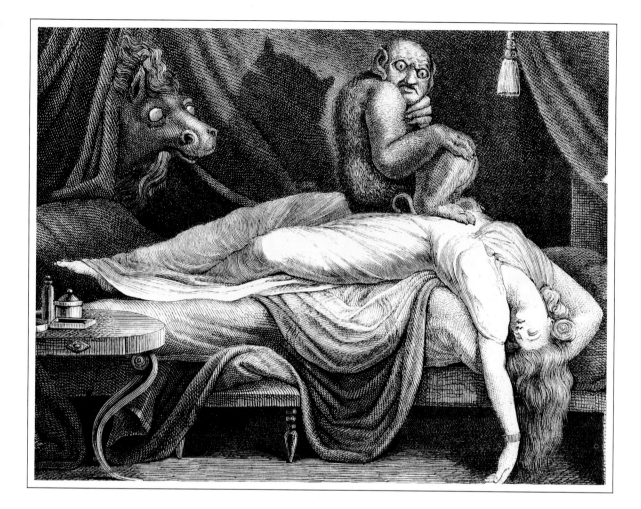

brothers 14 percent, and sisters 12 percent. Most dreams are straight stories without emotion, with perhaps the dullest on record belonging to a Midwestern college student who was recorded as having a dream in which he saw nothing but his forearms and the top of a lawnmower, and heard nothing but the buzzing noise of the lawnmower as he guided it over an infinite lawn.

Beyond here there may be monsters — Fuseli drawing of what dreams may beget.

When there is emotion in dreams, the dreamer is more likely to be worried about bad things than be happy about good. Fear leads at 40 percent, anger is felt in 18 percent of dreams that have emotion, happiness in 18 percent, and sadness in 6 percent. Consistent with this, hunted animals such as the rabbit and deer dream for only 15 or 20 minutes in a day, while hunter animals like the lion, bear, and even ordinary house cat can dream for over 3 hours a day. Outlines are clear in dreams, visual imagery predominates (except for the long-term blind, who dream more in sounds and smells), and black and white dreams seem common, though this might be an artifact from surveys before NBC heralded the way to all-color television in the late 1960s under prompting from the manufacturers.

This odd mix of sights and personalities is one of the most crucial things our bodies can experience. Interrupting someone's dreams all night will make him act drunk or punch-happy the next day or two and would probably be fatal if kept up on normal subjects for over two weeks. Tests of sleep and dream deprivation always have to be brought to an end before that time, because the bodies of the volunteer subjects start to weaken catastrophically, as at the very end of an ultimate stress response. As always though some bodies are different, and Mr. Leslie Gamble of Durham, in the north of England, has never slept since being bumped on the head in 1970 by a forklift truck at Heathrow airport. At bedtime he crawls into the covers with his wife until she's asleep then comes down to read or wander across the moors until dawn. Gamble says he would give up everything he owns for a single good night's sleep.

For most people though interfering with dreams is one of the most effective methods of torture used by the modern specialists in that unlovely subject, and, in the general guise of sleep deprivation, has been used by the British army in Northern Ireland and the American army in Vietnam, as well as apparently almost every other country that feels itself in the advanced league.

Dream deprival seems to be especially effective against otherwise committed political prisoners, as the Soviet NKVD found with the Old Bolsheviks, the Nazi SS found with French Resistance heroes, and the American Green Berets have found with democrats and intellectuals in a vast number of poor countries around the globe.

The power of dreams has been recognized since ancient times, dream interpretation being frequent in the Old Testament, in Rome (Nero's famous dream of being covered with ants and fearing it would be his subjects turning on him), and very much in Arab countries, until Mohammed outlawed its use, ostensibly on religious grounds, though since he admitted that his own understanding of Islam as being the ultimate religion and his being its prophet came to him through a "revelation," it might be suspected that a desire to steer off the competition entered into his dictum.

Freud, of course, made dream interpretation respectable in this century, and curiously enough much of what he said holds up in his own dreams as well. At various points in his writings he records over twenty of his own dreams. Many of them deal with receiving friendship, which might suggest an encouragement anxiety, something supported by the fact that he was certainly not the least prickly of men to work with, and would easily become worked up whenever someone did not promptly answer one of his letters. He also dreamed a lot about oral imagery such as food, eating, restaurants and the like, which might suggest a fixation on the presexual stage, but seeing how he satisfied oral longings in daily life — the famous cigar — perhaps it was not presexual after all. That hits at the weak point of his theory, for everything in the world that might be dreamed of is either convex or concave, and so can be interpreted in a sexually suggestive fashion. But everything which is convex is also concave, if you consider its obverse, so the reasoning can always go both ways, as in the good doctor's cigar being inserted or insertable, thus reducing the clarity of its meaning. Mrs. Freud's observations would undoubtedly be instructive but were, alas, never elicited.

What is happening in our brains to produce this dream state? One is that the cells used during thinking are no longer in the deep, almost knocked out stage they were during deep sleep. They're back to flickering in individual bursts, almost as if awake, and brain temperature and blood flow are back up from the low, cold levels at which they were during deep sleep. You're likely to sweat: an average of one pint comes out each night. The underlying recovery processes that went on during the deepest sleep stage have ended, including rebuilding of arm muscle, refreshing the inside of the leg bones, producing the sex hormone testosterone (it's especially high in deep sleep) and the like. It almost seems as if the body is awake, except for the key fact that during dreams it's more immobile than at any other time of day or night. In fact, it's frozen.

Shortly before the night's first dreams start, most people turn over to their favorite sleeping position, and stay locked there for the ten to fifteen minutes of dreams that is to come. Considering the possible contortions of the human body, it's surprising that the number of sleeping positions people lock into before dreams is so limited, for almost everyone chooses one of the big five, which are: lying on the stomach, said by at least one survey to be common in people who overorganize their days; lying on the side with feet crossed, correlating with a fear of deep relationships: lying on the side tightly tucked, said to go with desire for protection; lying on the back with the head resting on interlinked palms, going with pseudointellectuality; and, finally, lying on the back, arms and legs spread out slightly, correlating with self-confidence and security. The last position is the most common one for snoring, which certainly does seem to go with blissful self-assurance. Whichever position is chosen, is retained for the duration of the dream.

That's unusual because at other times during the night there is much tossing and turning, with six rollings-over and repositionings an hour being the minimum below which even the most sluggish of heavy sleepers do not descend. What causes the totally loose immobility of dreaming is yet another cluster of cells in the brain, this one located about a half-inch up from the reticular formation, near the main fluid-filled hole (or ventricle) inside the brain, and forming a mottled streak like an ultraminiaturized elliptical galaxy. Just before your dreams begin the clustered cells there squeeze tight and squelch out a burst of noradrenaline molecules, shaped, as we've seen, almost exactly like the potent adrenaline, except that they lack small bumps of carbon and hydrogen on one side.

The molecules float toward a nerve center in the brainstem that stretches down into the spinal cord, and makes those projections halt action in separate leads stretching out to the body's muscles. The result is that almost all the muscles you could otherwise so easily move during the day are now paralyzed into total immobility. This happens every time before you dream, and when it doesn't happen you don't dream.

Henry Kissinger taking an unscheduled rest.

One not very inspiring consequence is that as you dream you start to drool, for the jaw muscles are no exception to the general muscular loosening produced by the stuff streaming from the miniature galaxy-shaped object in the bottom part of the brain. You can see this phenomenon in someone who starts to fall asleep sitting in a chair. At first his eyes close and he breathes deeply, but he's still able to hold his head up. Such might happen without too much attention at business conferences or around the table at a heavy dinner. But as the suddenly snatched sleep progresses, the time for dreams comes quickly, and the galaxy-like structure does its stuff. Out comes its chemical, the brain cells controlling the voluntary muscles get blocked, the neck muscles go soft, and the head slides down. That slide, often preceded by a wobble and a few bobs until the chemical fully spreads through the motor sections of the brain, is what gives the ostensibly attentive sleeping one away.

When dignifiedly lying down in more acceptable nighttime sleep, or daytime siestas, there's no place for the pillow-resting head to fall to. But even then the jaw-loosening is not without its consequences.

The relaxing mouth muscles provide a mammoth-sized exit for the constantly secreted storehouses of saliva normally kept sanitarily tucked within. Through the suddenly enlarged orifice it dribbles without hesitation, and that's why the pillow of anyone awakened from a dream is likely to be so discomfitingly wet.

Daydreaming is more discreet in the salivary capacity, but it still shares with regular dreams the focus on insecurity. One study has shown that blacks, Italian Americans and Jews daydream a lot more often than Americans of Irish, German, or English background, a distinction which if valid, might tie in with relative entrenchment in the mainstream of American society. A particularly high number of daydreams are reported taking place on the bus.

In the general loosening into immobility of the body's muscles during dreams, there are two most notable exceptions. One is the eyes. They dart back and forth during a dream, tugging strenuously back on the tiny muscles that string them to the eye socket and pull them along. Why this happens is unclear, for blind people who dream in sound images will sometimes get this hyper eye twisting despite not seeing a thing. If it's just to track the images presented during a dream, one wonders why the calm ordinary speed tracking of daily eye motion is not enough. A tennis match is the only occasion when most people will consciously move their eyes this quickly back and forth, but except for the odd McEnroe or other fuzzy ball fetishist it's doubtful whether tennile subjects predominate in most people's dreams. Still, one extraordinary use has been made of this little understood eye excitement. It has been made to serve as a way of communication from out of a dream. An explanation is in order.

Many of us have at one time or another experienced the sensation of dreaming, and of being able to control or direct the dream. In other words, we were somewhat conscious during the dream. That experience has always had to rest as anecdotal, for in a paralyzed body, which is what all dreamers are in after the tiny elliptic galaxy streak has had its way, there's no way to talk or write about one of these semiconscious dreams as it happens. Or so it seemed. But since the eye muscles escape the general paralysis, one British researcher at the University of Hull decided that they could be used by someone who's in control of his dream to signal what's going on.

In one impressive experiment, the researcher, Dr. Keith Hearne, asked a number of subjects to try to communicate out of the next conscious dream they might have. The arrangement in his laboratory was as follows: if a flying scene occurred during one of these dreams in which the subjects felt aware, they were to flick their eyes quickly four times in the same direction just before flying, and five times afterward.

The researcher wired up his subjects in the laboratory putting tiny pickups on the sides of their eyes to detect the hoped-for movement of the muscles there, and brain-wave monitors on their heads to tell when they were dreaming. When the brain waves showed that one of the sleepers was dreaming, the researcher and his colleagues would stare at the eye movement readout to see if the message of four flashes and then five appeared. When they did show up, the sleeper was quickly awakened to see if that really had been a conscious message willed out from his dream.

It was! Most of the time the sleeper said that yes, he had been dreaming, had been dreaming of flying, and had been trying to get across the four- and five-flick message that would let Hearne and his crew know it. What's more, Hearne found that by giving the sleepers light electric shocks on the wrist when they were dreaming, he could often trigger them into those special dreams that they could control. With a bit more development, a perfected device along these lines might be marketed so each of us could produce lots of these deliciously controlled dream fantasies, otherwise so rare in us.

That's what might come from the eyes' being an exception to the general paralysis during dreaming. What might come from the other exception is perhaps better known for during dreams the sex organs enter into operation with exuberance. The vagina moistens and the penis grows erect. A lot. Studies show that most men have erections for some 95 percent of their dream time, which, as dreams occur cylically through the night, translates into an average of five erections lasting twenty-one minutes each — every night. This well-nigh miraculous leavening power of dreams (occurring almost equally in adolescents, adults, and even those otherwise impotent) happens during all dreams even those not overtly sexual.

No one knows quite why, though there has been no dearth of investigators eager to examine the subject in more detail. In the cause of science, penises of sleep-laboratory volunteers have been surrounded with doughnut rings of water, pasted with electrically sensitive skin-temperature monitors, and lassoed with steel thread like strain gauges to measure size. In one published experiment, subjects were required to sleep naked under clear plastic sheets, while patient and scientifically pure observers were enlisted to observe through a window the concerned member's development, judge it on a 0–4 scale, and if the sleeping position (remember the poor prone overorganizers) made long-distance viewing unclear, they were to enter the bedroom and peer up close for a good look. What the breakfast-time conversation in the laboratory was when the sleepers met their observers has been lost to history.

Further Rhythms and Sleep

Once the first dream period of the night has ended, slack in jaw if not genitals, your sleeping body redescends the levels as if it were just going to sleep again, arriving in another hour or less at the deep-sleep level from which the rise to dreaming occurred before. Down and up the cycle goes, five or so times in a night. The lightest moments of the sleep schedule are when you're most likely to be woken by a passing car, or a chance grunt by your mate. It's quite frequent to wake up at those 1½-hour intervals, peer around befuddledly, make feeble gurgling noises, then fall back asleep without even remembering that break.

You usually have to be awake a full three minutes or more in the middle of the night before your alternately paralyzed, stimulated, and

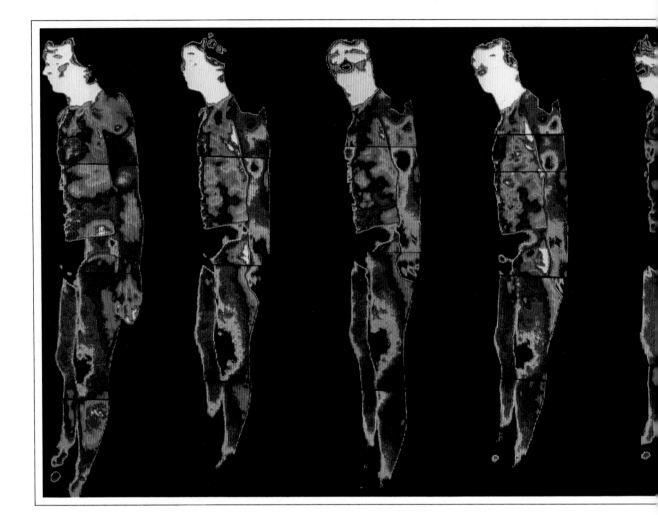

sexed-up sleepy brain will be able to remember it. When older people report starting to wake up a lot in the night, having short, interrupted sleep, it's most often not that they're awakening any more frequently than when young, but that they're staying awake long enough for that three-minute consciousness bar to be passed.

The up and down sleep schedule is so regular that you can tell a lot about a sleeper just from how long it's been since he went to bed.

Wake someone an hour into a nap and you're likely to reach him just as he's entering the deepest point, and most unlikely to be pleased by the intrusion. An insomniac who gets even two hours of sleep in the night is certain to feel better for it in the morning, for he will at least have managed to get through one complete cycle. Roll over back to sleep in the morning instead of getting up at the usual time, and you're likely to almost instantly swoop back into the world of your dreams, for the normal sleep period ends with a long, steady plateau of constant dreaming. This, incidentally, is why flying westward usually produces a less disagreeable jet lag than flying east: the number of dreams doubles in the first third of sleep after an eight-hour westward

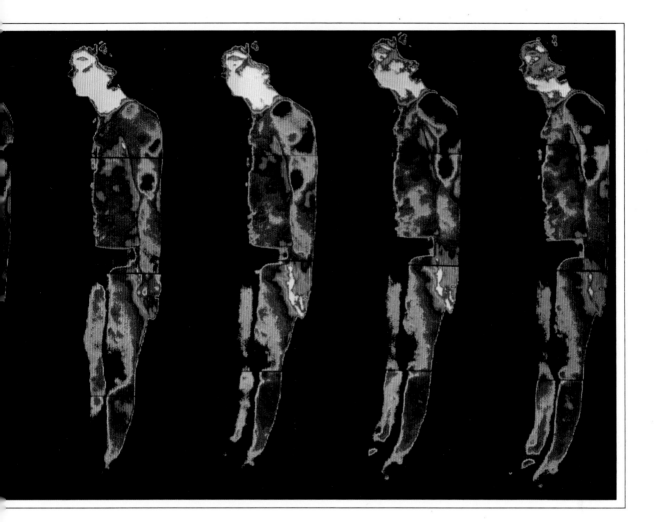

Dreams showing on the skin. In these heat images of a sleeping man wearing only briefs taken at half-hour intervals from midnight to 4 A.M., hand and face temperature varies in regular cycle. White shows skin temperature above 95, purple is 92, and black is under 89 degrees. Note that shoulder cools off when face does near end of sequence; chest heats up when face reaches maximum temperature at 1:30; knees are always coldest. None of these changes would be visible to the naked eye; dreams are most likely to have been at maximal heat points of midnight, 1:30 and 3:00 A.M. Room temperature was just 63 degrees.

flight — what would be morning to the traveler — on the eastward flight there's no such chance.

In the typical seven and a half hours of sleep a night 60 percent of the time will have been spent in light sleep, 18 percent in deep sleep, and 22 percent in dreams. The cyclic division shows up only after infancy, with preschoolers often having no light sleep but going directly to deep sleep and then dreams, while fetuses, at least those of seven months, have been measured in the womb as dreaming 84 percent of the time. (Knowing what the fetuses dream, would be fascinating, but what they think the rest of the time might be even more interesting.)

The light/deep/dream cycle goes on of course within the larger twenty-four-hour cycle that the body most readily lives in. It's common to feel more energetic in the middle of the day than late at night, and the body is very much behind this sensation. Your temperature reaches its lowest point between 3 and 5 A.M. almost 2 degrees Fahrenheit under the day's normal, and sleep is almost irresistible then. It's no coincidence that the 1979 accident at the Three Mile Island nuclear plant took place at 4 A.M. Going up and down in time with the day are

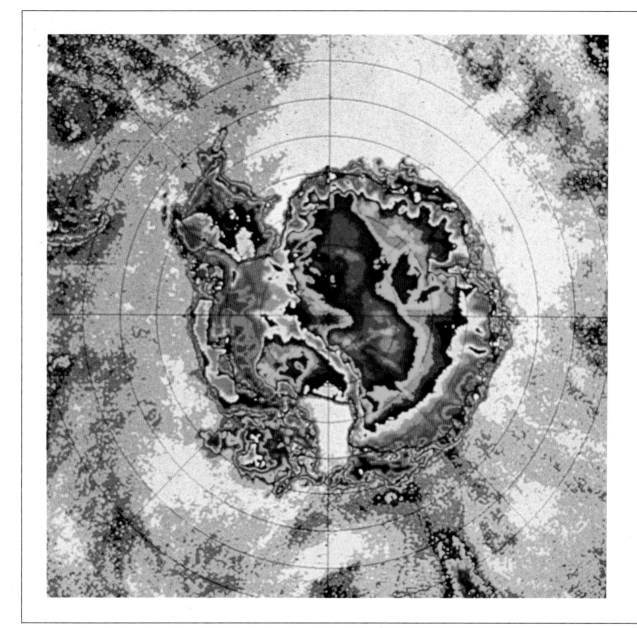

blood sugar levels, temperature, tear production, urine production, skin growth, and even liver enzyme production. This happens whether you follow the feelings they produce and go to sleep or not.

When Alexander Haig, then U.S. Secretary of State, was shuttling twice daily between Buenos Aires and London in his attempts to head off the Falklands War, he was exposing his body to the worst that unsettling of human biorhythms can do. The poor man undoubtedly suffered from diarrhea, salivary disorders, headaches, disorientation, problems in urinating completely and possibly blocked sinuses as he went through the meetings with junta leaders and Margaret Thatcher;

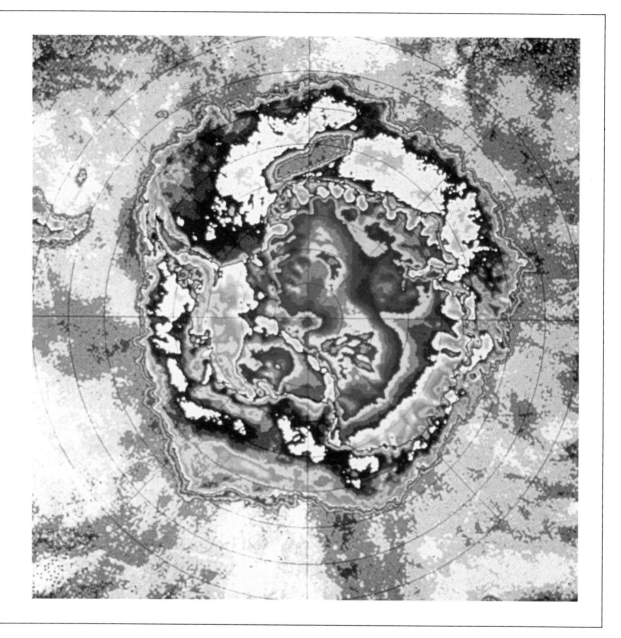

Antarctic winter icecap, in satellite survey. Icecap buckling due to moon's gravity produces vast slowing friction on earth's rotation, which over eons has helped create present 24-hour day length, and resultant evolution of human biorhythms — including evening tiredness and morning freshness — to that schedule.

the failure of this and other frantic diplomatic shuttle missions is quite likely linked to this deterioration of the too hastily traveling principal party. If you're awake at 4 A.M. up studying, carousing, shuttling or whatever your blood sugar and all the other cyclically changing substances are at their low point (deaths from infectious maladies peak then), while eight or nine hours later they'll be back up at their high just in time to get you to full steam by lunch. They sink down again after lunch, which is what's behind the midafternoon feeling of tiredness that has been countered in different cultures by the siesta, the habit of five o'clock tea (invented by the Duchess of Bedford around

1775 to keep her from "getting so tired"), and what was called *hnap-pian* in King Alfred's time, and is now the "nap."

The reason we have a twenty-four-hour tired/not-tired schedule is that it puts us in harmony with the sun, and safely out of the way come night. Certainly this schedule would be useful for hunted creatures, and indeed levels of blood hormones that change on the twenty-four-hour schedule are especially abundant in the tiny pineal gland, which in humans lies in the very middle of the brain, and in lizards is exposed to the changing rays of the sun through a clear slit window on the back of its head, suggesting an original connection with the day/night cycle.

To really see why our chief biorhythms go on a twenty-four-hour schedule, it's necessary to see why the pattern of day and night, which we have evolved to follow, takes twenty-four hours too. The answer is that it hasn't always. Four hundred million years ago the day was only twenty-two hours long, and a creature that automatically followed the cycle of light and dark then would have its hormones, blood sugars and the like oscillating on a twenty-two-hour cycle. The reason it's shorter now is the moon. Going around the earth it produces tides, and the sloshing of whole oceans up and down, even if it's only a few feet, puts a lot of friction on the ocean floors. That friction slows the earth's turning.

The moon's pull also produces tides in the semisolid Antarctic ice-cap, making it buckle up and down, like a child flexing a circular metal cake-tin cover. Each bend is just a matter of inches, but it's there, and measurable, and that mile-thick sheet of ice flexing over a good-sized continent produces even more slowing. So, if you wish to be pedantic the reason you get tired at night is because the Antarctic ice sheet shimmies under the power of the moon, which gives an added cosmic significance to bleary eyes and droopy head that prefigure the sleep that, when finished, will have you lying on your bed, ready to sit up and start another day.

Moonrise and sunset, Paradise Bay, Antarctica — where evening sleep begins and ends.

Illustration Acknowledgments

The author is grateful for permission to use illustrations from the following sources:

Animals Animals: 18 (Stephen Dalton), 217 (Leonard Lee Rue III)
Biophoto Associates: 6 bottom, 31 bottom, 50, 106, 145, 178, 258
British Museum (Natural History): 132, 202
C.E.T.I.N. (Centre Technique des Industries Mécaniques): 6 top
Charing Cross Hospital Medical School: 131, 142
C.N.R.I. (Centre International de Recherches Iconographiques): vi, 20, 25, 34, center, 34 bottom, 54, 58, 59, 111, 121, 137, 146, 150, 163, 164 bottom, 188, 195, 196, 200, 211, 213, 240, 247 (all photos in series), 257, 266, 270, 271
Bruce Coleman Limited: 21, 95
Ellis and Lacey: 190
Mary Evans Picture Library: 276
French Atomic Energy Research Unit, Orsay, France: 160 (Raynaud), 206 (Hasson), 261 (Comar)
Gower Scientific Photos: 13 bottom, 87 top, 87 center, 144
Gruppo Editoriale Fabbri/Milano: 3, 13 top, 40, 49, 60, 81, 88, 90, 209
Magnum: 17 (Bryan Campbell), 34 top, 62 upper left (Henri Cartier-Bresson), 159 (Eve Arnold), 230 (Leonard Freed), 250 (© Thomas Höpker), 279 (© Marina Faust)
Prof. M. Mukarami: 101
Lennart Nilsson: 109
O.N.E.R.A. (Office Nationale Etudes et Recherches Aérospaciales): vii–ix
Picker International, Hammersmith Hospital: 2, 46
Photo Researchers, Inc.: 98
Science Photo Library: xii, 8, 10, 11 (Dr. Ray Clark), 31 top, 38, 42 (Eric Gravé), 47, 62 top right, 62 bottom right, 64, 66 (Nancy Moorcraft), 76 (Dr. Ray Clark/Mervyn Goff), 78 (David Parker), 87 bottom (Dr. G. Schatten), 92, 104, 114, 127, 129 (John Walsh), 135, 139, 170, 172

Index